W9-BFL-168

Finish Carpentry

Fine Homebuilding®

BUILDER'S LIBRARY

Finish Carpentry

The Taunton Press

Library of Congress Cataloging-in-Publication Data

Finish carpentry.
p. cm. – (Builder's Library)
"29 articles...from past issues of Fine homebuilding" – Introd.
At head of title: Fine homebuilding.
Published simultaneously in paperback as: Fine homebuilding on finish carpentry.
"A Fine homebuilding book" – T.p. verso
Includes index.
ISBN 1-56158-064-3 (hardcover)
1. Finish carpentry – Miscellanea. I. Fine homebuilding.
II. Series.
TH5640.F56 1993 93-9974
694'.6 – dc20 CIP

Taunton
BOOKS & VIDEOS
for fellow enthusiasts

Cover photo: Kevin Ireton

First printing: June 1993

Printed in the United States of America.

A Fine Homebuilding Book

The Taunton Press, Inc.
63 South Main Street
Box 5506
Newtown, Connecticut
06470-5506

CONTENTS

INTRODUCTION

Details. That's what finish carpentry is all about. It's demanding and precise work, but it's really what stands out and separates the memorable from the merely acceptable. In fact, in many people's minds, finish carpentry is what determines value.

The 29 articles in this volume from past issues of *Fine Homebuilding* demonstrate the magazine's attention to detail—and the attention to detail required of careful builders, whether they're professionals working on new construction or home owners looking to make improvements. The articles cover a wide range of finishing subjects from installing crown molding to running baseboard efficiently, with special attention to the use of major tools and techniques involved in such finish work. The articles are written by well-known professionals and presented with the clarity of expression that readers of *Fine Homebuilding* have come to expect.

—Jon Miller, Associate Publisher

A footnote with each article tells you when it was originally published. Product availability, suppliers' addresses and prices may have changed since the article first appeared.

Installing Crown Molding

Upside down and backwards is the secret

by Tom Law

The first piece of molding (left) is cut square and run into the corner. The second piece (right) is cut to the shape of the molding's profile (coped) and will butt neatly into the face of the other piece. The paper-thin point on the bottom of the coped piece will make the finished joint look like a miter.

The old-time carpenters I learned from used to amuse themselves by quizzing young apprentices about the trade. If you could answer the easy questions, the last question would always be: "How do you cut crown molding?" And when you looked puzzled, they'd go off, chuckling to themselves something about "upside down and backwards." Of all the different moldings, crown molding is the most difficult to install, largely because of how confusing it can be to cope an inside-corner joint.

In classical architecture, crown molding (sometimes called cornice molding) is the uppermost element in the cornice, literally crowning the frieze and architrave. These moldings were functional parts of the building exterior when the ancient Greeks used them, but they have been used for centuries on interiors purely as decoration.

Crown molding is installed at the intersection of the wall and ceiling. Originally crown molding was triangular in cross section—the portions abutting the wall and ceiling formed two sides of a right triangle, and the molded face was the hypotenuse. But only the molded face is visible, so much of the solid back has been eliminated to save material. Also, by eliminating part of the back, only two small portions of the molding bear on the wall and ceiling surfaces, which makes crown easier to fit to walls and ceilings that aren't straight or that don't form perfect right angles.

The crown tools—When I cut crown, I like to work right in the room where the molding will go so I can orient myself to the wall I'm working on. If a room is finished, however, I may have to do the cutting somewhere else. Then I have to imagine the molding in place when I'm positioning it in the miter box (and believe me, this can get tricky with crown molding).

I cut and install crown molding with hand tools. I use a wood miter box (top left photo, p. 10) because it's the kind I learned on, but also because my view is not obstructed by the electric motor of a power miter box. Installing crown molding is slow and calls for careful work, so the production speed of an electric miter box is not required. I cut miters with a standard 26-in. handsaw (10 or 11 point). Miter cuts are made through the face of the molding, and a sharp handsaw will do a better job than a dull circular-saw blade will do.

For this kind of work, I prefer a workbench to a sawhorse. Mine is just a simple frame of 2x4s and 1x4s with a 2x12 top. It stands 34 in. high, which is a more convenient height to work on than a sawhorse provides. You don't need to deliver a lot of power to cut trim. A broad bench top is also convenient for holding tools.

Although you can still find deep-throated coping saws in the mail-order catalogs, most coping saws nowadays are 5 in. deep and have a 6-in. blade. The blades come with different numbers of teeth. I try each kind of blade to see which works best with the wood I'm cutting. Generally, finer teeth work best with hardwood and coarse teeth with soft wood, but not always. For the job shown here, I used a fine-tooth blade to cut soft wood. It works more slowly, but makes a smooth cut.

The blade of a coping saw can be inserted with the teeth directed toward you to cut on the pull stroke, or away from you to cut on the push stroke. Although it's strictly a matter of

From *Fine Homebuilding* magazine (February 1989) 51:64-67

personal preference, I orient mine to cut on the push stroke because it acts in the same manner as a handsaw.

Measuring and marking—Crown molding can be used by itself or combined with other moldings, but it should always be in proportion to the size and height of the room in which it's installed. Too much molding at the ceiling line tends to lower the ceiling visually. Three or four inches of molding at the ceiling line is about right for an average-size room.

In the rooms shown in this article, I used 3⅝-in. crown molding, which is the most common size available. This dimension is the total width of the molding, but it's not the critical dimension that you use when installing crown. You need to know the distance from the intersection of the wall and ceiling to the front of the molding, measured along the ceiling. Because the back part of the molding has been eliminated, you can't measure this directly. Instead, I put the molding inside a framing square to form a triangle and read the distance (top drawing, right). The molding shown here measures 2$\frac{1}{16}$ in. I mark that distance on the ceiling at each corner of the room and in several places along the walls. These marks will serve as a guide when I install the molding.

It's frustrating to drive a nail through a piece of molding and not hit anything more solid than drywall, so I locate the framing members ahead of time—when I can still make probe holes in the wall that will be hidden by the molding. If the room hasn't been painted, you can spot the studs and ceiling joists from the lines of joint compound and make pencil marks on the wall to guide the nailing. If the room has been painted, you can find the studs and joists by tapping with a hammer and testing with a nail. Electrical outlets and switches are nailed into the sides of studs and offer a clue to stud locations.

Running crown molding should be one of the final jobs on a new house. Walls and molding should be primed and first-coated, the molding installed and then finish coats applied. If you're retrofitting crown molding, it should be prefinished entirely so that all you need to do is touch up the paint or stain after installation.

On this job the walls were painted and the molding was prestained, so I marked the stud locations right on the crown molding. Rather than use a pencil to mark the wood, I made a slight hole with the point of a nail, which was easier to find in the dark stain and which I later nailed through.

Getting started—When I run crown molding in a typical room—four walls, no outside corners—I usually start with the wall opposite the door (bottom drawing). Unless it's perfect, a coped joint looks better from one side (looking toward the piece that was butted) than it does from the other (looking toward the piece that was coped). By first installing the crown molding on the wall opposite the door, and coping the molding into it on both ends, the two most visible joints show their best side to anyone entering the room.

I put the first piece up full length with square cuts on both ends. I cut it for a close fit, but if it's a little short I don't worry. Any small gap will be covered by the coped end of the intersecting piece. I hold the molding in place, lining up the top edge on the 2$\frac{1}{16}$-in. mark, and nail it. I use the shortest finish nails that will reach the framing, usually 6d or 8d. I nail into the wall studs through the flat section of the molding near the bottom, and I nail into the ceiling joists or blocking through the end of the curve near the top. I don't nail too close to the ends when I first put the piece up. I leave them loose to allow for a little alignment with the intersecting piece.

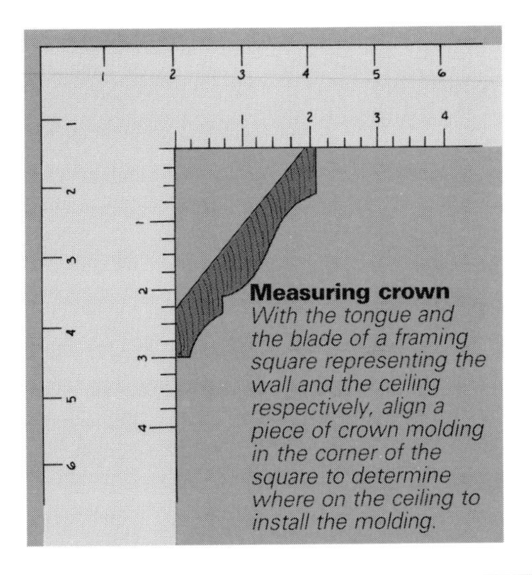

Measuring crown
With the tongue and the blade of a framing square representing the wall and the ceiling respectively, align a piece of crown molding in the corner of the square to determine where on the ceiling to install the molding.

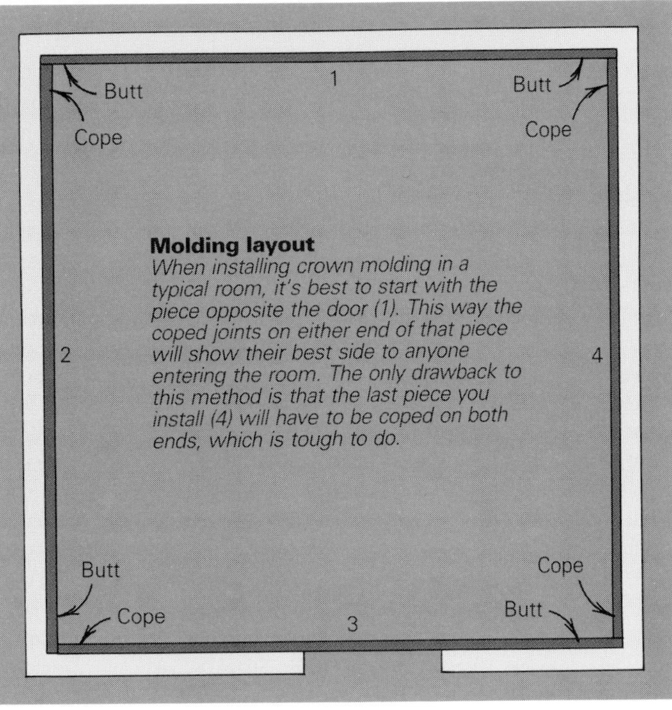

Molding layout
When installing crown molding in a typical room, it's best to start with the piece opposite the door (1). This way the coped joints on either end of that piece will show their best side to anyone entering the room. The only drawback to this method is that the last piece you install (4) will have to be coped on both ends, which is tough to do.

Occasionally I need to pull the top edge of the crown tight against the ceiling, but there's no ceiling joist or blocking to nail into. When that happens, I use 16d finish nails to reach all the way to the double top plate on the wall. Or sometimes I put a little glue on the molding and drive a pair of 6d finish nails at converging angles into the drywall about ½ in. apart (drawing next page). This pins the molding to the drywall while the glue dries. Another trick for nailing up crown when you don't have adequate framing is to nail up triangular blocks as shown in the drawing on p. 11.

I usually work around the room from right to left because I'm right-handed, and making coped joints on the right is a little easier than on the left. The second piece of crown to go up needs to be coped on the right and square cut on the left.

Coping with crown—Finish carpentry can be harder at times than cabinet work because you often have to make perfect joints against imperfect surfaces. Coping inside joints rather than mitering them is one way to deal with that problem. If you miter the inside corner with crown molding, the joint will often open up when you nail the pieces because the wall gives a little. A coped joint on crown molding won't open up and will be tight even when the walls are not exactly 90° to each other.

A coped joint is like a butt joint, with one piece cut to fit the profile of the other (photo facing page). The first piece of molding is cut square and run into the corner. The second, or coped piece is made by cutting a compound miter on the end to expose the profile of the molding, then sawing away the back part of the stock with a coping saw, leaving only the profile. The end of the coped piece will then butt neatly into the face of the first piece.

Because I don't always get the cope right the first time, I start with a piece of molding longer than I need and cope the end before cutting it to length. The phrase "upside down and backwards" refers to the position of the crown molding in the miter box when you're coping it (photo next page, top left). Crown molding isn't laid flat against the side or bottom of the miter box; it's propped at an angle between the two, just as it will be when installed. But the edge that will go against the ceiling is placed on the bottom of the miter box and is therefore "upside down." The right-hand side (if that's the coped end) is placed on the left and is "backwards."

The crown has to be positioned so that the narrow flat sections on the back of the molding, which will bear against the wall and ceiling, are square against the side and bottom of the miter box. When they are, the bottom should measure out the

Drawings: Michael Mandarano

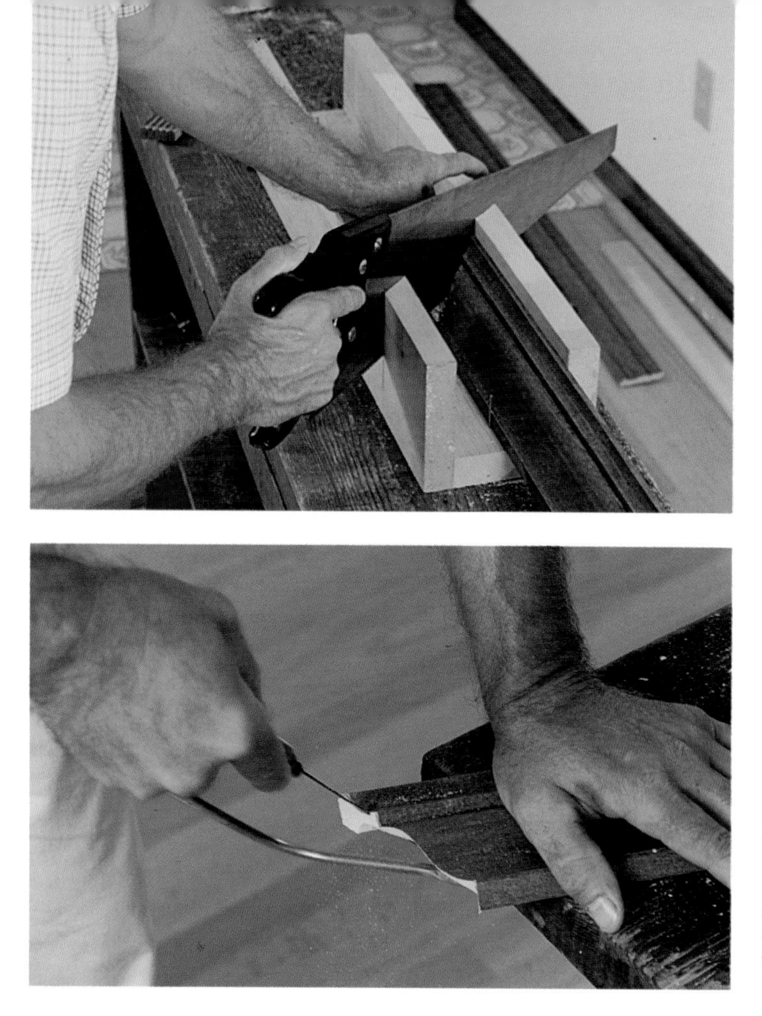

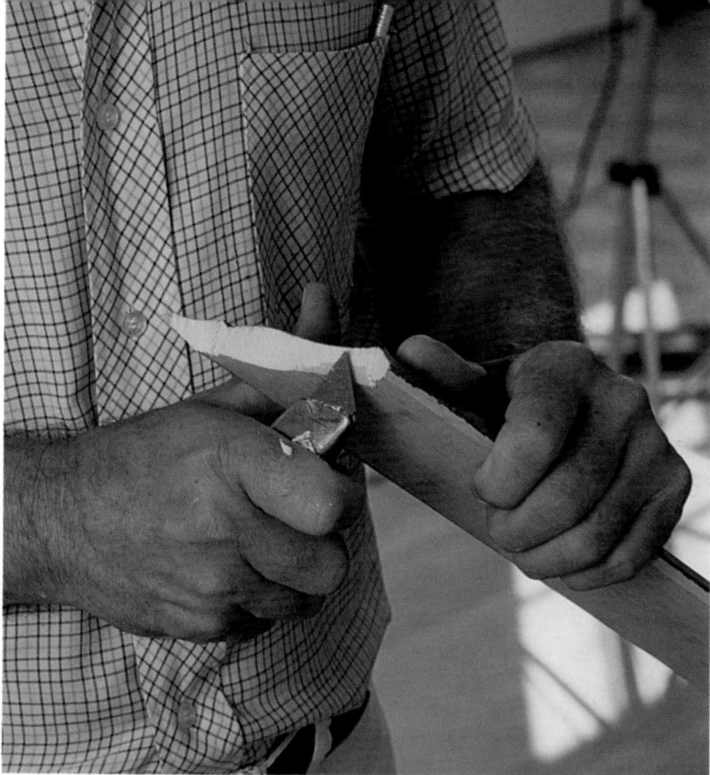

Before the crown molding can be coped, the end must be mitered to expose the profile. Positioned "upside down and backwards" in the miter box, the molding rests against small nails that hold it at the proper angle (top left). After exposing the profile of the crown molding, the back part of the stock is cut away with a coping saw, which must be held at a severe angle or the coped joint will not be tight (left). Even with the coping saw cutting at a severe angle, it's tough to remove enough wood through the S-curve in crown molding. Additional stock often has to be pared away with a utility knife (above).

required 2½₁₆ in. Once I find this position, I usually draw a pencil line on the bottom of the miter box to help me position subsequent pieces. Sometimes I'll even put a few nails on the line or glue a strip of wood to it.

Even with the molding positioned correctly in the miter box, it's still easy to cut it wrong. When I make the 45° cut to expose the profile of the molding for the cope, I remind myself that I want to cut the piece so the end grain will be visible to me as I look at it in the miter box (photo top left).

Coped joints are always undercut slightly, but crown molding has to be heavily undercut through the S-curve portion of the crown (called the *cyma recta)* or it will not fit right. I start the cope at the top of the molding using light, controlled push strokes. If I'm having trouble going from the straight cut to the curve, I back the saw out and come in at a different angle to cut away the waste. I begin the curved line with a heavy undercut and hold this angle all way through. I cut as close to the profile line as I can (photo above left).

The bottom of the crown molding is made up of a horizontal flat section, a cove and a vertical flat section. I cut down to the upper flat and then take the saw out and start cutting from the bottom. Some carpenters simply square off the bottom, but I try to leave the little triangular piece intact (photo, p. 8). I support it with my thumb as I'm coping and slice it paper thin. This little piece makes the coped joint look like a miter and helps close any small gap if the first piece didn't fit tightly to the wall.

I always test the cope against a scrap piece of molding to make sure I'm in the ballpark before actually trying it in place. Despite my best efforts to undercut the curved section, I usually have to pare away some more wood with my utility knife (photo top right).

I cut the piece just a little long and test it in place before cutting it to final length. If the fit of the coped joint is close, but still a little off, I can sometimes improve the fit by twisting both pieces either up or down the wall at this point—the 2½₁₆-in. mark on the ceiling isn't sacred. The buildup of spackle or plaster in corners can distort the intersection of wall and ceiling. Some carpenters carry a small half-round file with them to fine tune the fit of the cope.

Around the room—Once the coped joint fits, it's time to cut the piece to length. You can measure the total distance from wall to wall, but I find it easier to measure from either of the two vertical flat sections on the molding that the coped piece will butt into. If I'm working alone, I either step off the measurement with a measuring stick (a 12-ft. ripping, for instance), or I'll drive a nail into the wall (above the line of the crown molding) and hook the end of my tape measure over it. Wherever I measure from on the wall, I'm careful to measure to the same place on the piece I'm cutting.

When the coped piece is cut to length, I nail it up just like the first piece, leaving the square-cut end unnailed for the time being. If I need to draw the coped joint tighter, I nail through the coped piece into the piece it abuts.

The third piece of crown molding goes up just like the second, but the fourth one needs to be coped on both ends (bottom drawing previous page), assuming the wall is short enough to be covered with a single piece of molding. I cut this piece about ⅟₁₆ in. longer than the actual measurement, bow out the middle, fit the ends and snap it into place. The extra length helps to close the joints.

Some carpenters don't like having to cope the last piece on both ends because there's very little margin for error. The way to avoid this goes all the way back to the first piece of crown molding that's installed. Rather than put up the first piece with square cuts on both ends, you can temporarily nail up a short piece of crown molding and cope the first piece into it (photo left, facing page). Then take down the short piece, work on around the room and slip the butt end of the

Nailing tip

When there's no joist or blocking to nail into, you can put some glue behind the molding, then drive a pair of 6d nails at converging angles through the molding into the drywall. This will hold the molding while the glue cures. The drawing below also shows where to place the nails when installing crown.

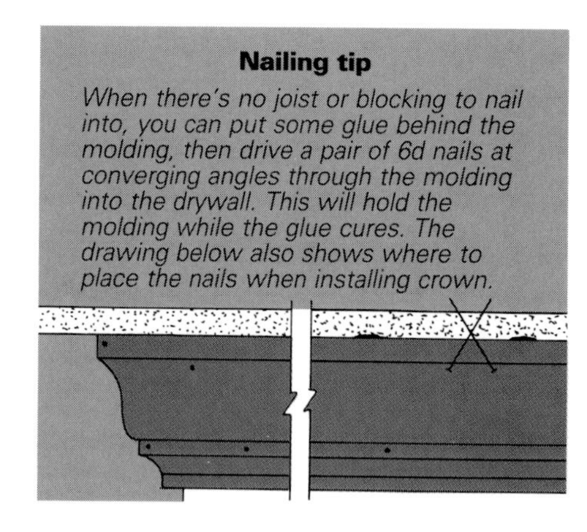

If you install the first piece of crown molding in a room by cutting both ends square, the last piece will have to be coped on both ends. To avoid this, you can put up a short piece temporarily and cope the first piece of crown molding into it (above). If an outside miter is open just slightly, sometimes you can close it by burnishing the corner with a nail set (top right). When a line of crown molding has to be neatly terminated on an open wall, the end should be mitered and "returned" into the wall with a small piece of molding. To avoid splitting such a delicate piece, it's best simply to glue it in place (right).

last piece behind the first cope that you made. This way all four pieces of crown molding in the room will have one square-cut end and one coped end.

When I go into a room that's not a simple rectangle, the decision about where to start is influenced by where I'll end. If there is an outside corner in the room, I like to end by installing the shortest piece that has an outside miter. That way, there's less wood wasted if I cut it too short. If there's not an outside corner, I like to work so that the last piece is installed on the longest wall that can still be done with a single length of molding.

When I need more than one piece to reach from corner to corner, I cut the moldings square and simply butt them together rather than use scarf joints or bevel joints. Butt joints are easier to make for one thing. And for another, although wood isn't supposed to shrink in length, the truth is it does. Over the years, I've seen a lot of joints that have opened up, and of those, the butt joints looked better than the others.

Outside corners—These are also mitered with the molding upside down and backwards in the miter box, but the saw is angled to bevel the piece in the opposite direction. When you miter for a cope, you expose the molding's end grain, but with a mitered outside corner, the end grain is *behind* the finished edge. Sometimes I cut them at an angle slightly greater than 45° to ensure that the outside edges mate perfectly. I usually add a little white glue, then nail through the miter, top and bottom, from both sides.

Sometimes outside corners will close tightly but the leading edge of one piece overhangs the other, perhaps because the corner is not exactly 90° or because one piece of molding is thicker than the other (more about that in a minute). If the molding hasn't been painted or stained, I'll trim the overhanging edge with a sharp chisel and sand it. This actually leaves a narrow line of end grain exposed at the outside corner, but once the molding is stained or painted, the end grain isn't very obtrusive. There are times when the molding has the finish coat already on it, and I can't do this because it would expose raw wood. In that case, I use my nail set to burnish the projection smooth (photo top right).

On this house, I ran crown molding in the foyer and had to terminate the molding at the stairwell opening. I ran the molding through the dining room, turned the corner at the stair and ended the molding with a return—a mitered piece that caps the end of the molding. To make a return, I simply cut a miter for an outside corner on the end of a scrap of molding, then lay the piece face down on the bottom of the miter box and cut off the end. I glue this in place with white glue so as not to take a chance on splitting it by using a nail or brad (photo above).

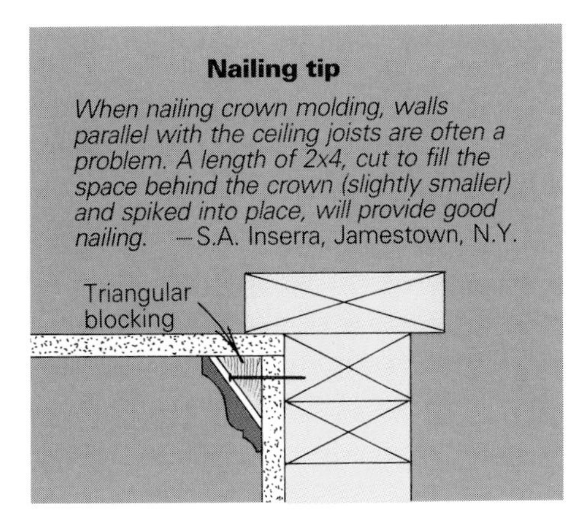

What can go wrong?—Whether because the wood was wet when it was milled, or because the knives were dull, or because of internal stresses in the wood, the exact dimension and profile of the pieces in a given bundle of stock molding varies considerably. The differences aren't obvious until you try to fit an inside or an outside corner with two pieces that don't match. It's best to make joints from the same piece whenever possible.

There are times when the wall or ceiling is so crooked that gaps are left along the length of the crown. If there is a short hump that causes gaps on each side, I scribe the molding and plane it for a better fit. If the gaps aren't too bad, it may be best to fill them with caulk. Another trick I've used is that of leaving a small space (usually about ¼ in.) between the top of the molding and the ceiling, which makes it harder for the eye to pick up irregularities. If I'm doing this, I put up blocks to nail to, as shown in the drawing at left, and use a ¼-in. spacer block to ensure a uniform reveal. □

Consulting editor Tom Law is a carpenter and builder in Frizzellburg, Maryland.

Installing Two-Piece Crown

A method for running wide, paint-grade crown moldings

by Dale F. Mosher

I work as a finish carpenter on the San Francisco Peninsula, where there is a resurgent interest in formal houses that have a Renaissance European flavor. The houses often have a full complement of related molding profiles for base, casings and crown, and to be in scale with the rest of the building, these profiles can be quite wide. In the case of the crowns, I'm talking 10 in. to 12 in. wide. In fact, the crown moldings that I sometimes install are so wide they come in two pieces (photo at right).

There are several reasons for making crown in two sections. First, the machines that cut the moldings typically have an 8-in. maximum capacity. There is a lot of waste when wide moldings are carved out of a single piece of stock. For example, I'd need a 3x12 to mill a 10-in. wide piece of crown—an expensive, inefficient use of the resource. Two-piece crowns are also a little more forgiving during installation. The type I used on the job shown here can be overlapped in and out a bit, allowing the width of the crown to grow and shrink as needed to account for dips and wows in the walls and ceilings.

Joined at the shadowline. **Wide crowns, which appear to be made of a single molding, can be made by running related profiles adjacent to one another. Any gaps between the two are caulked prior to painting.**

All the two-piece crown moldings I've encountered have been custom-made. The designer or architect comes up with section drawings; then the mill shop has the molding-cutter knives cut accordingly. Here, an average set of custom knives costs $35 per in., plus there is a $75 setup charge for each profile. So before the wood starts to pass over the cutters you've already spent a fair amount of money. But to create a certain look, it can be money well spent.

Stain-grade or paint-grade trim—If you've got one, your architect or designer will decide what grade the trim should be. If you don't, your checkbook will decide. At the mill, stain-grade means the stock is clear and virtually free of knots. On the wall, stain-grade means no opaque finishes will be applied. The moldings are individually scribe-fitted, and that means no caulks or putties to fill any gaps that may occur at miters and along uneven walls. Paint-grade material, on the other hand, will have some sapwood and uneven grain. On the wall, it can have caulkable gaps. Obviously, the stain-grade material will cost more—how much more depends on the species of wood. And it costs a *lot* more to install. Where I work, we figure four to five times more labor is needed to put up stain-grade crown as opposed to paint-grade.

The most commonly used paint-grade materials in these parts are alder, poplar and pine. I see more poplar than anything else because it's relatively inexpensive and easy to mill. Even paint-grade moldings, however, don't come cheap, and they should be handled with care. I have them primed on both sides as soon as they are delivered, and I store them on racks with supports no farther apart than 3 ft.

Test fit. **After affixing the top portion of the crown to its backing blocks, Mosher checks the fit of its corresponding bottom half.**

Miter-box station—I've used the new sliding compound-miter saws to cut crown, and I've decided to stick with my 15-in. Hitachi chopsaw. Here's why: When you're running crown, you've got to make both back cuts and bevel cuts. It takes time to adjust the saw back and forth, and the constant changes multiply the chances for error. Also, the crown has to lie flat on the table with a sliding saw, which makes it harder to see the path of the blade and the cut line. None of these is a problem when using a chopsaw.

The key to cutting crown accurately is having a good station for the miter box (top photo, next page). Mine has a pair of wing tables that flank the saw, connected by a ¾-in. MDF (medium density fiberboard) drop table that supports the saw. Each wing table is 6 ft. long and 2 ft. wide. They can easily be stood on end and carried through a standard doorway. I can also put a 2-ft. deep table against a wall and have enough clearance behind the saw to swing it through its settings.

The wing tables have fences that support backer plates for the crown during a cut. To

From *Fine Homebuilding* magazine (December 1991) 71:85-87

install them, I begin by snapping a line on the floor where the assembled station will sit. Then I put the outside legs of the tables on the line, and anchor each leg to the subfloor. Next I put the miter box in the drop section of the table. I removed the stock fences from the miter box, allowing me to bring the arbor slightly forward of the original fence line, thereby increasing the width of the cut. The wider of these two crowns was 7¼ in.—just about the limit of what my saw can handle.

I make sure the saw's turntable can swing freely from side to side, and that its blade is square with the tables when set at 0°. I secure the saw to the drop table with four drywall screws run through holes that I've drilled in the saw's base. Each wing table has an 8-in. high fence that's square to the blade. I align the fences with a string to make sure they're straight.

During a cut, the crown bears against ¾-in. MDF backer plates that are screwed to the tables and the fences. The backer plates should be ½ in. wider than the widest crown section. The crown moldings for this job meet the wall and ceiling at 45°, so I ripped the backer-plate edges at 45°. Other crowns meet the wall and ceiling at different angles, and the backer-plate edges should be beveled accordingly.

It took me about a day to build this setup, and another half day to fine-tune everything. But the time it takes to build one will be returned tenfold in a single good-sized installation job.

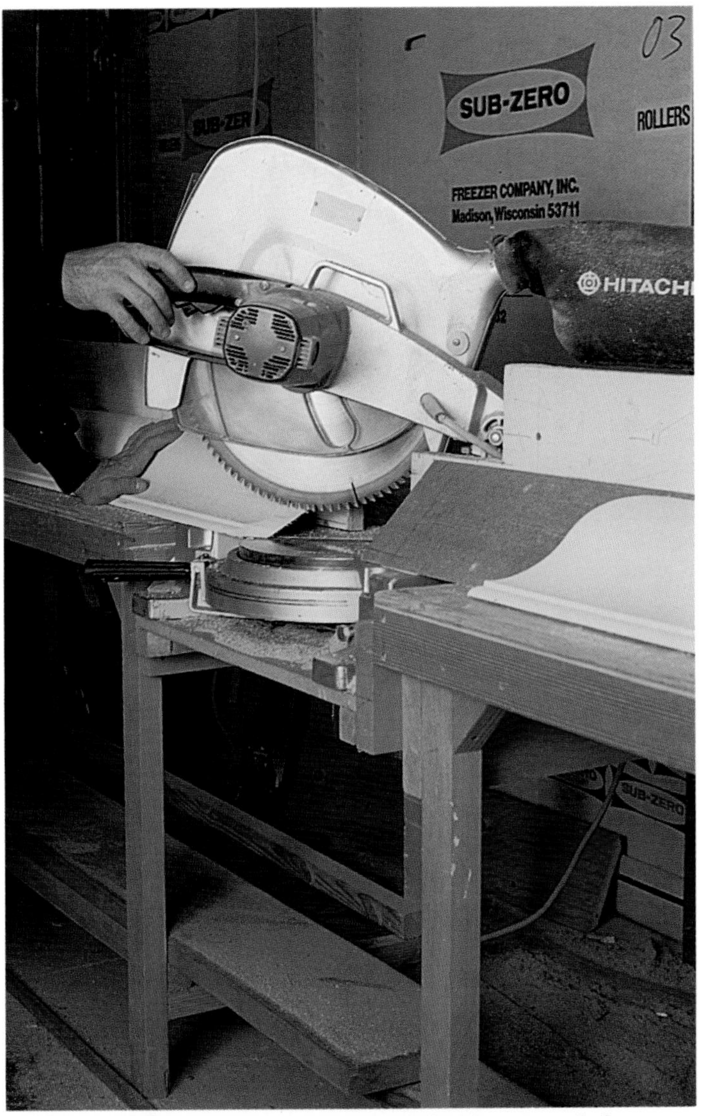

Miter-box station. **A pair of wing tables linked by a saw platform provides angled backer plates to support the crown moldings as they are cut. The plates are screwed to the table and fences.**

Backing blocks—A two-piece crown needs a solid base for nailing and a flat surface to rest against to ensure correct alignment for the pieces. Backing blocks serve this purpose (top left photo, facing page). To find the width of the backing block, I assemble a couple of short sections of crown, as they would appear when installed, and measure the backside from the point the assembly hits the ceiling to the point it engages the wall. The backing blocks should be 1/16 in. less than this measurement to ensure that the crown will go together without leaving gaps between the pieces, at the ceiling, or at the wall. Backing blocks can be made of solid wood, but I prefer ¾-in. plywood because it's affordable, doesn't split and it holds nails well. After ripping a stack of backing-block stock, I cut the blanks into 6-in. to 8-in. lengths.

I prefer to place backing blocks on 16-in. centers, and no farther apart than 24 in. They should be affixed to the framing, so if the painting crew is about ready to prime the walls, I mark stud and joist locations with a keel (a carpenter's crayon) along the ceiling/wall intersection. A keel will bleed through most primers. Omission of this step gets you a one-way ticket to the planet of frustration, where you poke nails into the walls and ceilings, looking for the lumber.

Dropped platform. **The wing tables are joined by a saw platform lowered far enough to bring the saw's table flush with the wings. The fences are braced from behind with triangular blocks on 1-ft. centers.**

Backing blocks are installed on layout lines snapped on the ceiling and wall. Using a torpedo level with a 45° bubble, I position a block at one end of the wall so the bubble reads level. The block should be about 2 ft. from the corner to avoid joint-compound buildup. I mark its edges on the ceiling and wall, and repeat the process at the opposite corner. These points are the registration marks for the chalkline. I use the raised chalkline as a straightedge to locate dips or bumps in the ceiling and walls. These problems are usually due to framing irregularities, and the backing blocks should be kept away from these places. I wish the framing crew could be around during this part of the job. If they only knew the trouble *we* go through to make *them* look good, they'd be taking trim carpenters to lunch a lot.

To attach the backing blocks I use a 2¼-in. finish nailer because finish-nail heads are small enough to be consistently set below the face of the block. Nail heads that stand proud interfere with the crown. When I've got framing on one end of the block for nailing, but none on the other, I put a bead of glue on the backing block to help anchor it to the wall.

At inside corners, I run one block into the corner, and then scribe the adjoining one to it. The backing blocks are typically a little too wide to fit between the lines in the corners because of joint compound on the wall, and they need to be trimmed a bit to fit. Outside corners are sometimes mitered as though crown molding, and secured to the wall, ceiling and to each other through their mitered edges.

When the backing blocks are up, I cut my "tester blocks." These are typically 16-in. to 24-in. long pieces of the crown molding. They need to be long enough to reach from an inside corner to the closest midspan backing block. I cut three pairs of tester blocks with inside miters at both ends, and three pairs with outside miters. One set has 44° miters at each end, one has 45° miters and the third has 46° miters. You may ask, "why not cope the inside corners?" For one, the curved profile of the widest molding in this job meant that a coped corner would be very fragile. I've found that a glued inside miter on a paint-grade job—if the pieces are carefully fitted—yields first-rate results.

Backing blocks. **Crowns this wide need substantial backing to provide a consistent plane for aligning the two pieces and for adequate nailing. In the corner, one block extends to the wall while the other is scribe-fitted to it. Here, a test piece of crown is held in place. The pencil line along its point intersects the corner formed by the two blocks, marking the point from which the overall measurement for the crown will be taken.**

Prybar tweaking. **As the moldings are nailed home, a small prybar is useful for aligning the adjoining sections.**

Unit assembly. **Short sections of crown are best preassembled into a single piece. The pencil marks on the ceiling show the points from which the crown-length measurements were taken. On the right you can see a fully assembled run of crown.**

Running crown—Installing crown is not a solo operation. The job will go a lot faster and with greater accuracy if you've got a good helper. Working on the theory that a piece of trim can always be made shorter, we begin with the longest run in the room by tucking the pair of 45° test blocks into one of the corners, just the way the finished crown will fit. If the fit isn't acceptable, we try a 44° and a 46° block until the right combination turns up. It might be a pair of 44s. It doesn't matter. It is very important, however, that the line of the miter line up with the corner, whether it's an inside or outside miter.

Once we find the best fit, we make a pencil mark along the bottom of the block into the corner (photos above). This marks the point from which the overall measurement is taken.

I don't bother to cut the piece a little long and then shorten it by degrees to ease into the fit. My helper and I can measure it accurately, so I cut it to that length. Period. This saves a lot of climbing up and down the A-frame scaffolds we typically erect as work platforms.

We also use tester blocks at outside miters to determine the best angles. To get our measuring points for an outside-to-outside miter, we make pencil marks on the ceiling to note the long points. For an outside-to-inside miter, we mark the long point of the outside miter, and the heel cut at the inside miter.

I'll typically put four 1½-in. finish nails into each backing block. The nails should be placed where the painter can easily putty the nailheads. I don't put nails in a tight radius or too close to an inside corner. A small prybar can be useful for aligning the crowns during nailing (photo above left). I prefer the ones used by auto mechanics.

Back-beveling the miters can be useful on recalcitrant fits. A good tool for this is a 1⅛-in. belt sander. Its protruding belt makes it very maneuverable. If I need shims, I use pieces of manila folder. At each miter, I run a bead of yellow glue to ensure a sturdy joint.

Sometimes the crown has to work its way in short sections around a wing wall. In this case, I usually preassemble the pieces if they're shorter than about 12 in. I put the parts together with glue and a pneumatic brad nailer, let the glue set for 20 minutes and then place it as a unit (photo above right).

As we run the upper crown, we make notes in the corners that describe any special angles or back-beveling that it took to get a good fit. Nine times out of ten, the same cuts will work on the lower section of crown.

After the crown is up, the drywallers can float on any necessary topping compound to hide the bumps and bows in the ceiling and wall. If the wails are to be textured, this should be done after the crown is installed. Our painters use oil-base putty to fill the nail holes, and latex-based paintable caulk to make the joint between the two pieces of crown disappear.□

Dale F. Mosher is a carpenter who specializes in finish work in Palo Alto, California. Photos by Charles Miller.

Making Curved Crown Molding

Glue-laminated trim can be shaped on a table saw

by John La Torre, Jr.

As a carpenter, I spend most of my time at work swinging a hammer or wielding a saw. But I'm always looking for a chance to try a new technique. Recently, while I was touring a house under construction, owner Paul Kreutzfeldt showed me a straight piece of stock crown molding he planned to use for the kitchen ceiling, then looked up at a curved wall in the room and said, "That piece is going to be a bear."

"Yup," I answered, excited. "Mind if I give it a try?"

"Go right ahead," he said with a smile.

There are two basic approaches to making curved trim. You can glue several pieces of wood end-to-end and then cut the curve on a bandsaw, or you can laminate thin strips of wood around a curved form. If you use the first method and decide to stain the trim, the separate pieces may accept stain differently, and the joints usually show through. Laminated trim, on the other hand, is stronger than butt-joined trim and usually looks better when it's stained. Because Kreutzfeldt had yet to choose between staining and painting his crown molding, I decided to laminate it. As it turns out, Kreutzfeldt decided to paint it (photo above).

Making the bending form—The first step was to make a form that matched the curvature of the convex wall. Finished with drywall, the wall defined a 90° arc having a radius of about 24½ in. Unfortunately, the curve was far from perfect, wandering out of round by up to ⅜ in. That forced me to make a template for the form.

To create the template, I bandsawed a 24½-in. radius curve in a sheet of ⅛-in. tempered Masonite. I then held this template against the curved wall 3¾ in. from the ceiling, which is where the bottom of the crown molding would contact the wall. After scribing the template with a pencil compass, I trimmed it with a jigsaw for a snug fit.

Molding in the round. **Made with basic shop tools, the laminated crown molding seen above wraps around a slightly out-of-round convex wall, butting at both ends into straight, factory-made crown. Photo by Rich Miller.**

Making the bending form itself was the easy part. Back at my shop, I traced the outline of the template onto three pieces of ¾-in. plywood, cut them out and nailed them together, placing ¾-in. plywood spacers between layers to produce the proper thickness. Finally, on the back edge of the form, I bandsawed a series of steps parallel to the front edge to give the clamps good purchase (photos next page).

Preparing the stock—The next step was to prepare thin strips of wood for lamination. The factory-made crown molding I wanted to match was made of white pine, but I selected clear sugar pine for my molding. Sugar pine has a uniform straight grain, is easily bent without splitting, and well, that's what I had on hand.

I produced the laminating stock by resawing ⅞-in. by 4-in. wide boards, which produced thin boards that were $^{3}/_{16}$ in. thick. I used my bandsaw for ripping the boards into strips because its $^{1}/_{16}$-in. saw kerf wastes less wood than my table-saw blade. Then I ran the strips through a 10-in. bench planer to remove irregu-

From *Fine Homebuilding* magazine (April 1992) 74:79-81

Glue up. The laminating form consisted of three layers of ¾-in. plywood separated by ¾-in. plywood spacers (photo above). A series of steps bandsawn into the back of the form provided solid footing for an assortment of clamps. Extra strips of pine (photo below) placed against the molding stock helped distribute the clamping pressure.

Low-tech shaping. The molding was contoured by making a series of cuts on a table saw to remove the waste up to the layout line. A mark on the rip fence indexed the center of the saw blade, indicating the optimal spot for the molding to contact the fence during the cutting operation.

larities and to reduce the strips to a uniform thickness of ⅛ in.

Next, I traced the cross section of the factory molding on graph paper. Examining this profile, I decided to laminate the trim out of three different widths of sugar-pine stock to simplify the removal of waste from the laminated blank.

With the strips cut to width, I dry-clamped them to the form to identify and eliminate any problems before glue up. I discovered that keeping the strips aligned would be difficult. My solution was to trace a slightly oversized cross section of the three-step assembly on two scraps of plywood, cut the patterns out and then slip the scraps over opposite ends of the assembly. These simple jigs helped to prevent the wood strips from sliding around during glue up.

Shake it up—For laminating jobs, I like to use urea-formaldehyde glue, which starts as a tan-colored powder that must be mixed with water (for more on builders' adhesives, see *FHB* #65, pp. 40-45). I used to employ a stick or a rubber spatula for mixing but sometimes ended up with lumps of powder that wouldn't dissolve. Then I discovered a better method: Put the powder in a plastic container, add the correct amount of water and then shake the container in a circular motion. Surprisingly, mixing in this way is faster than with a stick and produces a lump-free mixture every time.

To make glue up simpler, I taped the pine strips together edge-to-edge on my glue-up table. Then I spread the glue across the assembly using a paint roller. That done, I removed the masking tape, coated the masked areas with glue, tilted the strips upright and pressed them together.

Glue up took every clamp I had (including C-clamps, bar clamps and pipe clamps), and it wasn't a pretty sight (photos facing page). I installed the first clamp at the midpoint of the form and worked my way toward both ends, alternating clamps above and below the form. Extra strips of wood placed against the outer plies of sugar pine helped distribute the clamping pressure evenly. Excess was scraped off before the glue cured.

After letting the glue cure for 24 hours, I removed the blank from the form. Checking the concave side of the blank against the template, I saw that the molding was within $\frac{1}{16}$ in. of a perfect fit. I decided not to shave it further just yet.

Sculpting on the table saw—Once the glue up was completed, I ran the curved blank, top-edge-down, through the thickness planer to remove slight irregularities from the bottom edge of the blank. Then I flipped the blank over and planed its top edge to size. Planing the curved blank was easy—I simply steered it through the planer to keep it perpendicular to the cutterhead.

With the sizing completed, I squared both ends of the blank and traced the outline of the factory crown molding on one end. Now all I had to do was remove everything that didn't look like crown molding.

Probably the easiest way to make crown molding is to cut it on a shaper. Many cabinet shops nearby had a shaper, but none had a cutter that matched my molding. I could have ordered custom-made cutters, but that would have cost $300 to $400, difficult to justify for a one-off piece of trim.

I decided to shape the blank by making a series of table-saw cuts to remove most of the waste (photo facing page). This worked remarkably well. I clamped a 3¾-in. tall board to the fence to make it the same height as the molding, then marked the top of the fence to index the centerline of the saw blade. While making each cut, I held the blank against the fence at the index mark. Any deviation from this mark was insignificant, because it merely caused the saw blade to wander into the waste area, requiring nothing more than a second pass across the table saw to get it right.

My blade cut a ⅛-in. wide kerf, so I made the cuts by moving the fence toward the blade in $\frac{1}{16}$-in. increments, raising the blade just enough each time to remove the maximum amount of stock without cutting across the layout line. In this fashion, I finished with a cross section very close to that of the factory crown.

Because the stock laid flat on the saw table, the cutting operation was accomplished safely and easily. I also kept my fingers far away from the sawblade at all times.

Scraping it smooth—All that remained was to smooth out the small, sharp steps on the molding blank. I figured this would be the easy part, but it turned out to be the most difficult.

First I tried sanding, but the sandpaper quickly became clogged with pine resin. I soon realized I'd have to scrape the pine smooth.

To make a scraper, I cut a 45° angle on the end of a scrap piece of the factory molding and traced its profile on the blade of an old taping knife. Then I cut the blade to the layout line using a bench grinder and raised a cutting edge by rubbing the blade with a hardened-steel punch.

Dragging this homemade scraper along the molding at a 45° angle produced satisfying results (photo below). Each pass left behind a small trail of fine dust instead of the curled shavings a perfectly tuned scraper would produce, but little by little the sharp steps began to disappear. Scraping the molding down to the layout line took two hours and a lot of elbow grease.

The scraping left the molding with some torn fibers and minor irregularities, so I decided to finish the job by sanding. I began by using 80-grit sandpaper to work out the irregularities, then worked my way up to finer-grit paper. I use Wetordry TRI-M-ITE sandpaper (3M Construction Markets Department, 3M Center, Bldg. 225-4S-08, St. Paul, Minn. 55144; 612-736-7761) because its backing doesn't tear while sanding. The sanding took about three hours and just about wore me out.

Installation—Earlier, when trimming the ends of the curved blank, I had left the ends 1½ in. long. Now I used the bandsaw to cut a 1½-in. long triangular stub tenon on both ends of the molding. These tenons would fit into the triangular voids behind the factory crown molding, aligning the joints while providing solid backing for the ends of the factory molding.

The final step was to fit and attach the curved crown to the wall. Using a belt sander, I relieved the concealed edges a bit so that the molding fit snugly against the wall and the ceiling. By now Kreutzfeldt had decided to paint the crown molding, but just a small amount of belt-sanding produced such a tight fit against the wall and the ceiling that no caulking was necessary. I applied construction adhesive along the back and top of the curved crown, then fastened the molding to the wall with screws run through the stubs.

As Kreutzfeldt installed the straight runs of crown molding, I was gratified to see that just a bit of sanding produced a satisfying match of curved to straight molding. □

John La Torre, Jr., is a carpenter in Tuolumne, Calif. Photos by author except where noted.

***A homemade scraper.* The author smoothed the sawcuts using a scraper made out of an old taping knife. The edge of the scraper was shaped with a bench grinder.**

Making Curved Casing

Strip-laminating arches to match straight casing profiles

by Jonathan F. Shafer

A few years ago I was asked to step in and complete the finish work on an 11,500-sq. ft. Tudor home that had taken 18 months to get through drywall. Completing the trim took an additional 12 months and posed many challenges, such as hanging 8-ft. high doors, building four stairways and running thousands of feet of wide casings and base. The house also had many arch-top windows (photo below) and doorways of various heights and widths. The curved window and door casings had to match the existing straight casing, so I decided to produce the curved casings on site with the help of a talented crew of finish carpenters.

My approach to this challenge was to strip-laminate the arched casings. By alternating strips from two pieces of straight, even-grained casing, we reproduced the casing profile. We ripped the strips from straight casing and then bent them around a form for each window and door. We also laminated extension jambs for each window, using the same bending forms.

Making the patterns—Our first step was to make patterns of all the arched windows and doors. How we produced the patterns varied depending on the particular application—some methods were as simple as tracing on kraft paper (available in long rolls) against the window frame, while others were as involved as mathematically computing arcs and multiple radius points.

One method we used on some of the more complex windows required a thin, flexible ripping of even-grained wood long enough to follow the arch along a window frame. This strip was clamped or held by helpers against the inside of the frame. We maintained the arch shape by tacking crosspieces to the bowed strip. The more crosspieces we used, the better the shape was held after the clamps were removed. We then transferred the shape of the arch to kraft paper.

With another method, we tacked plywood against the window, using a piece wide enough to contain the unknown radius points. We then used a beam compass to find the radius points on the plywood by trial and error. Again, the arch was then transferred to kraft paper. I came on the project too late to have done it, but in the future I would make tracings of each window frame prior to installation.

Finally, we cut out each pattern and checked it against the corresponding window, making necessary adjustments. The patterns also had to be extended on both ends to allow extra casing length for trimming later. We labeled the patterns for window location and wood species.

Building the bending forms—When the patterns were ready, we built a bending form for each one. We constructed them from 2x stock cut into arcs on a bandsaw (for the design of a bending form for curved jambs, see the sidebar on p. 21). With roundtop casings, the 2x arcs were made using a simple circle-cutting

The curved casing for these windows was fabricated on site, using two pieces of straight stock cut into narrow strips and laminated around a form.

From *Fine Homebuilding* magazine (April 1991) 67:82-8

jig fixed to the bandsaw table (drawing below). We extended the table with a piece of ¼-in. plywood and ran a screw through it to create a pivot point. The 2x stock was then pivoted around the pivot point on a ¼-in. plywood carriage.

To cut the more gradual arcs of the bigger windows, we used a 1x3 to extend the pivot point of the circle-cutting jig across the shop (photo below). The 2x arcs were screwed to a plywood base or to the subfloor, depending on how big they were.

Ripping strips—Once the bending forms were completed, the strip-cutting operation was next. The basic principle here is that you're taking a piece of straight casing with the molded profile you want and ripping it into narrow strips that you can bend around a form and glue back together. But if you were to do this by ripping a single piece of casing, the resulting molding would be narrower than the original because of the material lost to the saw kerf. Therefore, you have to make alternate cuts on two pieces of straight casing.

To ensure that the laminating strips were cut to a uniform width, we used thin pieces of pine as spacers resting against a preset table-saw fence. This enabled us to cut the casing incrementally without changing the position of the saw fence.

In our case, the saw blade, and hence the laminating strip, was roughly ⅛-in. wide. The spacers were cut so that each was twice the width of the table-saw blade. We cut our spacers the same length as the short auxiliary fence on my table saw. To keep them from slipping with the casing as it was being cut we simply tacked a brad to the underside of each spacer, which hooked over the front edge of the saw table (drawing next page).

After the spacers were completed, we adjusted both pieces of casing (ripped a little off them) so that the finished width was an even number multiple of the spacers. Our casing had a rabbeted back band around the outside edge, so we were able to reduce slightly the outside edge of the casing without changing the profile. (The side casings were also adjusted in width to make them equal to the arched head piece.)

Next, we glued and clamped the back band to the casing. We also filled in the plowed relief on the back of the casing by gluing in thin material and jointing it flush. This was necessary so that each strip would be cut square to the others.

We set the table-saw fence to equal the total adjusted casing width (casing plus back band) minus the width of one spacer. Finally, we equipped our table saw with a riving knife (or splitter) mounted behind the blade. This protected the thin strips from damage as they came off of the saw.

In order to produce alternating strips from two pieces of straight casings, we ripped the first piece of casing on the saw with the outside edge against the fence (drawing next page). Next, the second piece of casing was ripped with the inside edge against the fence, and so on.

On the first piece of casing that we did, I made the mistake of dropping the laminating strips into a pile, thinking I could easily sort them out later. Wrong. It took me over an hour to put the pieces in order before gluing them up. For subsequent casing, I built a rack to hold the strips in order (top photo, next page). The rack was simply a pair of 1x4s with saw kerfs in them, nailed to a short bench.

Glue up—Before we could start gluing, the bending forms had to be adjusted to allow for the jamb reveal—the difference between the inside edge of the jamb and the inside edge of the casing. We tacked ¼-in. thick spacing shims (the width of the reveal) against the forms. We also covered the forms with waxed paper to prevent the casing from adhering to them during glue up.

We had plenty of pre-adjusted clamps on hand to do the job—everything from bar clamps to wedges against wood blocks screwed to the floor. Our clamping cauls were bandsawn to match the outside radius of the casing. We dry-fit the strips around the form so that we could work out a clamping strategy (bottom photo, next page). During glue up, we quickly and evenly brushed yellow glue on each piece. Because the casing was relatively wide and the set-up time relatively short, we glued and clamped the strips in three stages and let the glue dry overnight before proceeding to the next stage. Once the complete casing was dry, we scraped and sanded the casing profile, removing glue squeeze-out and any irregularities in the profile.

Making extension jambs—We made the extension jambs for the windows with the same bending form used for the casing—all we had to do was remove the ¼-in. shims. The reveal was ¼ in. so we made the extension jambs ⅝ in. thick, allowing sufficient material to secure the casing. We produced strips roughly ⅛ in. thick to reduce the chance of springback and wide enough to fill the space between the window frame and the edge of the drywall, plus ¼ in. extra for ripping and jointing to the finished width after the glue had dried.

Ripping and jointing was a two-man operation. One man fed the piece into the table saw or jointer, and the other helped support the piece as it went into and came out of each machine. Cutting and fitting the extension jambs to proper length was a trial-and-error process, the error always being on the long side until the jambs fit. Next they were glued and nailed to the window frames through predrilled holes.

Fitting the casings—The ease of fitting the casings to the windows was directly related to the care with which the pattern had been made. If the pattern was true to the window form, the casing was relatively true to the window. Because our laminating strips were about

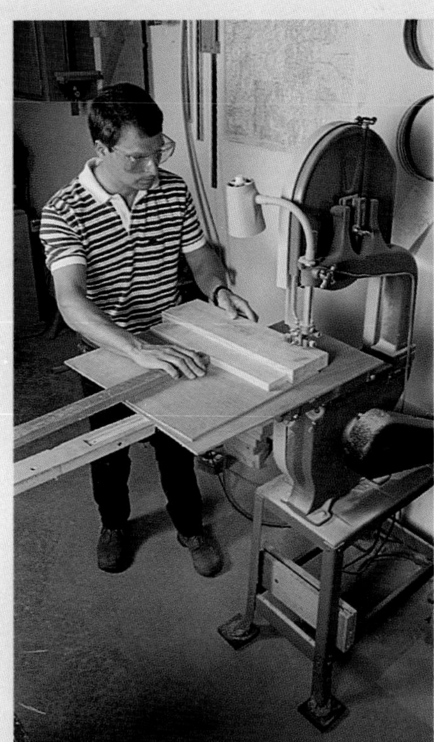

For round-top windows with small radii, Shafer cut sections of the bending forms on the simple jig shown in the drawing. For bigger windows, he attached a length of 1x3 to the plywood carriage (above) and extended it across the room to a center point on top of a workbench.

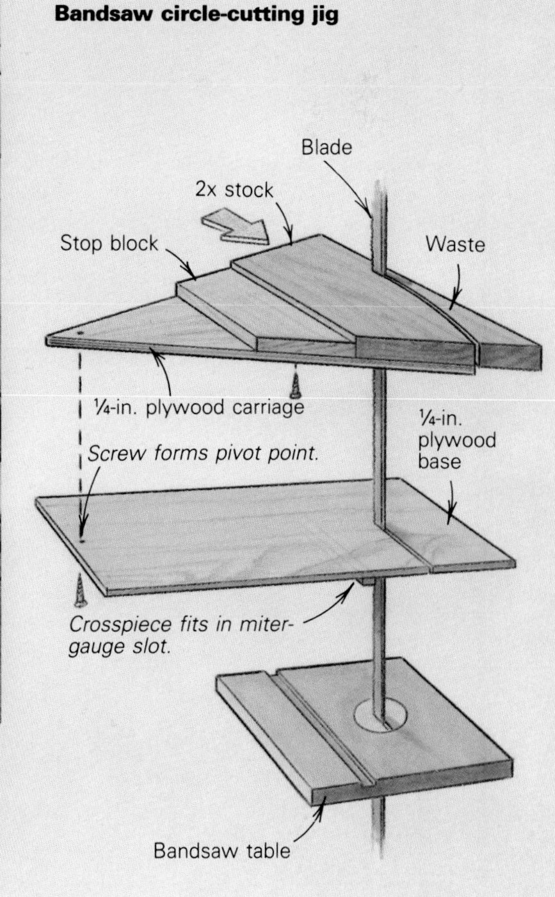

⅛ in. wide, the springback was negligible. We were using relatively wide casings, so springing the casing to match the window—anywhere the pattern was not true—was very difficult, if not impossible. We had to live with compromises in a few places.

Just as with the extension jambs, fitting the casing was a trial-and-error process. It was relatively simple on the windows with one-piece casings that were butted directly to the window stools. Likewise, using plinth blocks would have simplified fitting the casing on the bigger windows and doors. But we decided to miter the corners between the arched casing and the side casings.

We calculated the miter by tracing the head casing and side casings right on the drywall, then connecting the points where their inside edges and outside edges intersected. The head casing was cut to match this line, and the corresponding angle was then cut on the side pieces. After the miter was judged to be tight, the bottoms of the side pieces were marked and cut square to rest on the window stool, or mitered if the window was picture-framed. □

Jonathan F. Shafer was a carpenter in Dublin, Ohio, when he worked on this project. He has since relocated to Bellingham, Washington. Photos by Kevin Ireton.

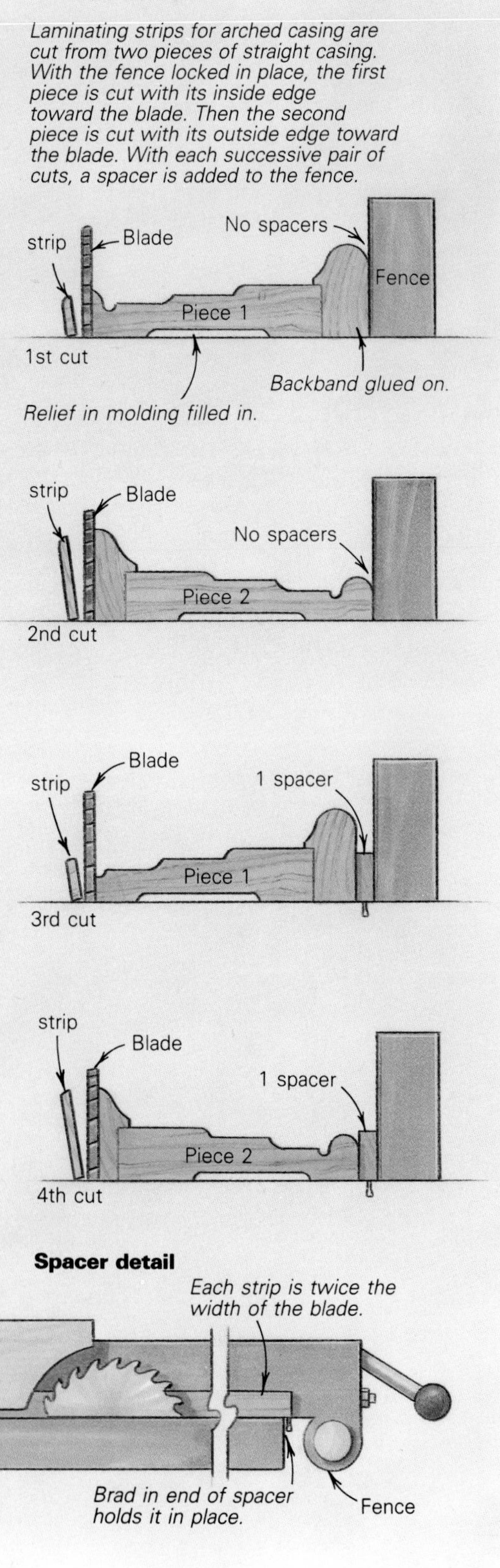

To ensure that the laminating strips were cut to a uniform width, Shafer kept the table-saw fence at a fixed distance from the blade and used spacer strips to move the straight stock incrementally closer to the blade. The rack beside the saw holds the laminating strips in their proper order for gluing.

The bending form is simply 2x stock cut into curved sections and screwed to a piece of plywood. Here the laminated strips have been clamped up without glue to work out the clamping strategy and eliminate some of the usual glue-up trauma.

Reusable jig for curved jambs

With the trend toward New Classicism in architecture has come a renewed popularity of arch-top windows and, to a lesser extent, doors. Making the jambs for these doors and windows presents only minor challenges. The millwork is straightforward, a matter of laminating a stack of plies in a curved jig. The problem is cost. One of the axioms in the millworking business is that no matter how many stock jigs you have cluttering up the back room, the next customer will want a window of a slightly different size. Building a custom jig for each order soon prices the work out of the market.

Over the last few years, David Marsaudon and the other folks at San Juan Wood Design have developed and refined a jig for making arch-top jambs that employs reusable parts. With this jig, they can lay up fair, smooth arcs of virtually any radius, and do so at a cost that gives them a competitive edge in the custom market. Their jig is made with some scrap plywood, strips of ¼-in. hardboard, a sheet of particleboard and a shop full of clamps. I visited San Juan's shop in Friday Harbor, Washington, and got a close look at how the jig works.

Building the jig—The jig, which somewhat resembles a dinosaur skeleton, is made of three parts: a semicircular particleboard panel, a set of clamping blocks and a mold surface. The radius of the particleboard panel determines the final size of the jamb.

I watched craftsmen Roger Paul and Jerry Mullis make the frame. They started with a given finished radius (the inside of the finished window frame) of 73 in. Based on previous trials, they estimated springback to be 2½ in., so the jig had to have a radius of 70½ in. By subtracting the thickness of the mold surface (¼ in.) and the depth of the clamping blocks (3 in.), they came up with a panel radius of 67¼ in. A new panel must be cut specifically for each size window frame, but smaller panels can be cut from larger ones to save on materials.

Paul and Mullis propped this panel vertically on a bench, and screwed to its rounded top a set of clamping blocks at about 8 in. o. c. Made from scraps of ¾-in. plywood laminated in pairs, these blocks are saved for reuse in each new jig. Finally, Paul and Mullis stapled a skin of ¼-in. hardboard, as wide as the intended window jamb, to the clamping blocks. They started at one end and carefully worked their way to the other, keeping the skin centered on the blocks and checking the final jig for bowing. Because any imperfections in the skin would telegraph through to the window jamb during glue up, Paul and Mullis checked that the skin was perfectly smooth. As a final touch, they waxed the skin with paraffin and smoothed the wax with steel wool. The wax keeps glue from sticking to the skin, so the finished jamb will pop easily out of the jig.

Assembling the plies—Curved jambs are built from four plies, each ³⁄₁₆ in. thick, about ¼ in. wider than finished width to allow for edge-jointing later, and a few inches longer than arc length. Paul and Mullis first sorted among the four plies and selected the best for the inside finished surface. They finish-sanded this face—a job that is much easier on a flat bench *before* glue up. They laid that ply face-down on the bench and on its back they drizzled aliphatic glue, each man releasing his artistic talents in a distinctive pattern of squiggles. They spread the glue to an even film, paying special attention to coating the surface along the edges. Then, they lifted the second ply onto the first and repeated the procedure until all four plies were glued. They clamped the plies by the edges to keep them together as a unit until clamped in the jig.

Clamping—Working quickly before the glue set, Paul and Mullis lifted the group of plies onto the jig and balanced it on the apex. Starting at the center, they laid a caul across the width of the plies, centered over a clamping block. They hooked one bar clamp under the chin of the block, snugged it down and then did the same with another clamp on the opposite side. Working as a team, one on each side of the jig, Paul and Mullis worked away from the center, clamping to each block with just enough pressure to hold everything in place. With the frame now held stable in the jig, Paul and Mullis went back and added two C-clamps, with a caul above and below, between each of the clamping blocks (drawing below). With all the clamps in place, they tightened them firmly, and glue flowed from every seam. They have found that only an even clamping pressure will give a fair arc. The key is to use lots of clamps and tighten them uniformly. The other men in the shop joked about Paul and Mullis hatching another peacock (photo below), and in fact, the radiating bar clamps did call to mind one of those strutting birds.

I returned the next morning to watch the plucking of the bird. Clamps were removed, scraped free of dried glue, and hung back in their racks. With the bench clear, Paul and Mullis laid the finished frame against the penciled layout marks. The frame had sprung to within ⅛ in. of the design width.

—J. Azevedo, a free-lance technical writer in Friday Harbor, Wash. Photo by the author.

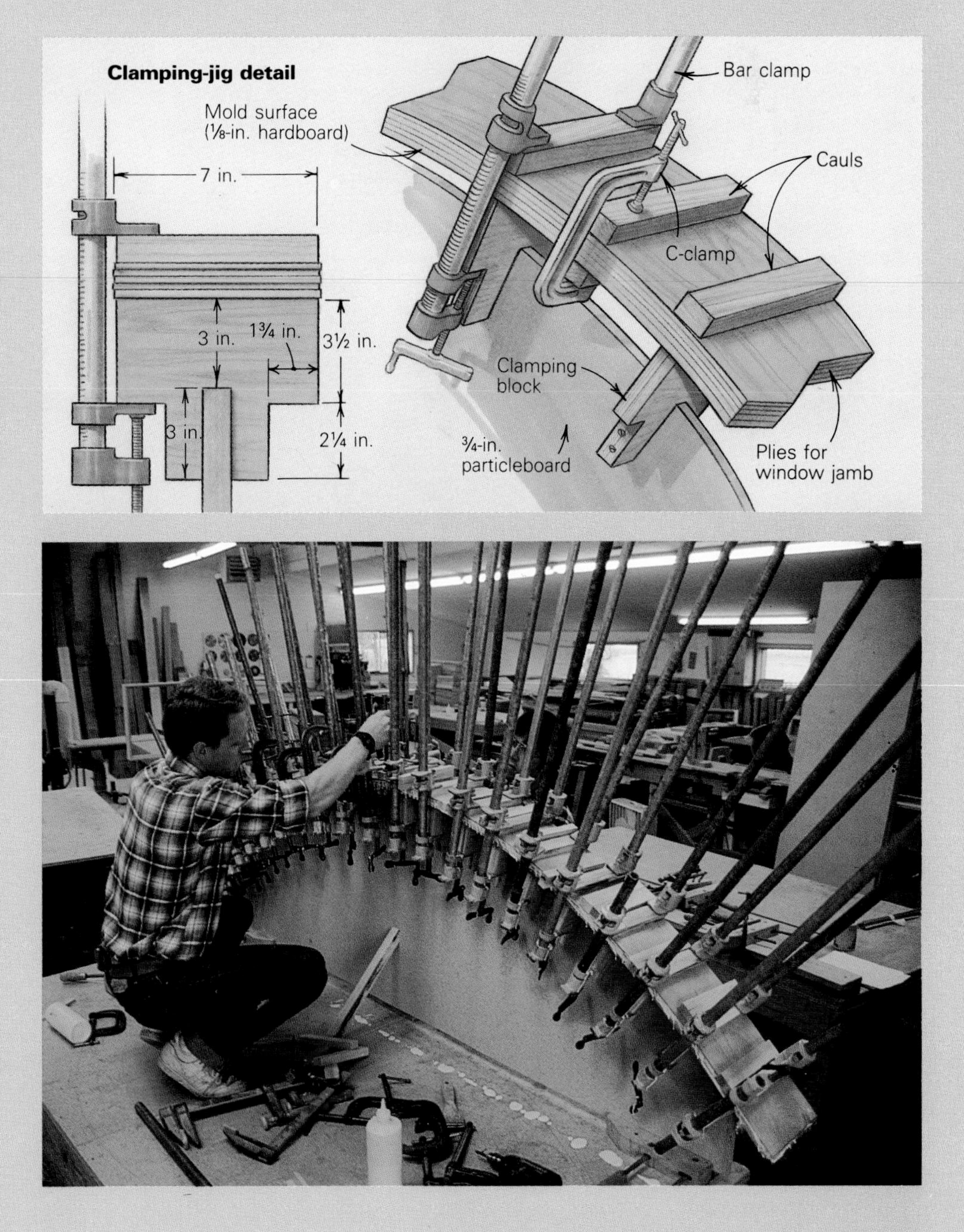

Cutting Crown Molding

Calculating miter and bevel angles so you can cut crown on compound-miter saws

by Stephen Nuding

A few years ago, I purchased an 8½-in. compound-miter saw. It was light and compact, but had the same capacity for cutting large crown moldings as a regular 10-in. miter saw. Remodeling Victorian homes, I install a lot of crown so this seemed to be the perfect power tool for me.

I eagerly brought the saw to the job and set the miter and bevel angles for 90° corners, as indicated by the instruction manual. When I cut my first lengths of crown, the joints weren't perfect, but I figured that the walls and ceiling weren't perfect either, so with a little shaving here and there I was in business.

The next crown molding I had to install, however, was a larger one, and when I cut it and held it up to the ceiling, I was looking at a pie-shaped gap ⅜ in. wide. What's more, this room had two corners that were 135°, not 90°, and the saw's instruction manual gave no miter or bevel angles for this situation. I soon discovered that throwing miscut pieces around the room in rage and frustration is a very slow and expensive way to complete a job.

By now I was ready to return the saw to the dealer and demand a refund. But in desperation I grabbed the instruction manual one last time. According to the manual, the miter and bevel angle settings were correct for 90° corners when using a standard crown, which makes a 38° angle to the wall. Wait a minute... what if my crown doesn't make a 38° angle to the wall?

Fortunately, my daughter's protractor was in the car, so I was able to measure the angle the crown made to the wall by holding it against the inside of a framing square. The angle was more like 43° or 44°. I checked all the crowns I was installing only to find that none were the same, varying from 35° to 45°.

I finished the day's work as best I could and went home determined to calculate the angle settings for each of the crowns. Using my wife's high school math text to brush up on some trigonometry, I wrote down equations and measurements. I worked late into the night, but couldn't come up with a formula.

I finished the crown job eventually by trial and error, playing with the angles on the saw until they were right. Still, the problem gnawed at me. I spent a lot of late nights scribbling and thinking, but I just couldn't get it.

Fortunately, I had hung some French doors in the home of Roger Pinkham, professor of mathematics at the Stevens Institute of Technology. So one Saturday morning, at my request, he graciously came to the house and we pored over my notes. Several hours later, we had it. We could calculate the miter and bevel angles for any crown and for any angle.

So, why use a compound-miter saw?—You are probably wondering why anyone would want to calculate angle settings for a compound-miter cut when crown molding can easily be cut on a regular miter saw with no math at all. With a regular miter saw, the crown is positioned at an angle between the fence and the table (photo below left), but is turned upside down so that the wall face of the crown is against the saw's fence and the ceiling face of the crown lies on the saw's table. The crown is then cut at 45° to create a 90° corner, 22.5° for a 45° corner, and so on. Very simple. (For more on cutting crown molding, see the article on pp. 8-10.)

Most 10-in. miter saws, however, can only cut crown molding up to about 4½ in. wide. Five and one half-in. crowns are readily available, though, and cutting these requires a 14-in. or 15-in. miter saw—a large, heavy tool. Cutting large crowns on any of these saws also requires the extra step of constructing a jig or fence extension, preferably both.

Even a 15-in. miter saw is not big enough to cut crown molding more than 6½-in. wide, and larger crowns are also available. For instance, the Empire Molding Co., Inc. (721-733 Monroe St., Hoboken, N. J. 07030; 201-659-3222) makes an 8¾-in. crown that I often use.

So unless you want to make a king-size

To miter crown with a standard miter saw, turn the molding upside down and set it at an angle between the fence and table.

To miter crown molding with a compound-miter saw, lay the molding flat on the saw's table.

From *Fine Homebuilding* magazine (June 1991) 68:79-80

miter box and cut the molding with a handsaw, you'll have to use one of the new slide compound-miter saws or a radial-arm saw to cut these wide crown moldings (for more on these saws see *FHB* #57, pp. 58-62). With a compound-miter saw, crown molding is laid flat on the saw table (photo right, previous page). No jig or fence extension is necessary. The saws can be smaller for cutting the same size crown, resulting in a lighter tool with a smaller blade, which is therefore cheaper to buy and costs less to sharpen.

Figuring the angles—To calculate the miter and bevel angles for any crown molding, you'll need a framing square and a calculator that's capable of doing trigonometric calculations. These calculators are usually called "scientific calculators." No cause for alarm, though, just think of yourself as a carpentry scientist. I use a Radio Shack model that is out of production now, but a Radio Shack EC 4008 will do nicely and retails for only $13.95.

So, here we go. First let's consider the most common case, the 90° corner. Hold whatever crown molding you're using up to the inside of a framing square as in the drawing right. Measure lines A, D and C to the nearest 16th of an inch. (To convert fractions of an inch to decimals, simply divide the denominator into the numerator. To convert ⅞, for instance, divide 8 into 7 and you get .875.) The miter-table setting (M in our equation), is the inverse tangent of (A divided by C).

$$M = \tan^{-1}(A \div C)$$

To calculate this, divide A by C, and then hit the inverse tangent button ($\tan^{-1}$), or arc tangent button (same thing). In our example, 2.875 (A) divided by 4.8125 (C) = .5974. With .5794 still on the calculator screen, hit the inverse tangent button and you get 30.9° (rounding to the nearest tenth of a degree). This is the miter angle at which to set your saw.

The bevel angle (B in our equation) is the inverse sine of D divided by (the square root of 2) times C.

$$B = \sin^{-1}\left(\frac{D}{\sqrt{2} \times C}\right)$$

To calculate this, multiply the square root of two (done on the calculator) times C. Then divide that into D and hit the inverse sine button, or arc sine (same thing) button on the calculator. Using the values from drawing A, the calculations would go like this: the square root of 2 = 1.41, times 4.8125 (C) equals 6.8059, divided into 3.875 (D) equals .5694, the inverse sine of which is 34.7°. This is the bevel angle at which to set your saw for a 90° corner.

Once you have calculated the miter and bevel angles for a particular molding, you never have to calculate them again as long as you have 90° corners. Jot down the angles somewhere and save a couple of minutes the next time you run that crown.

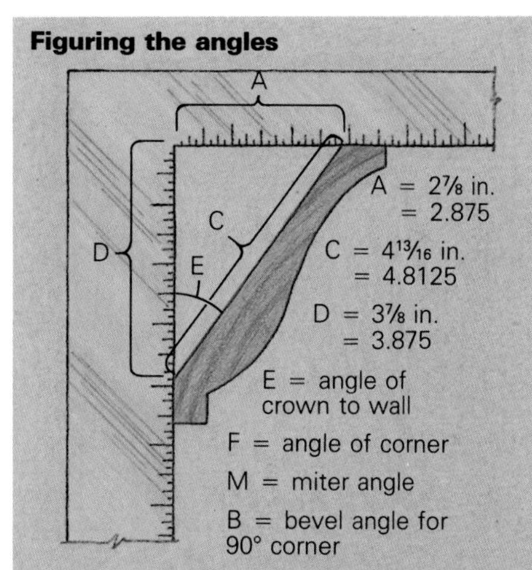

Crown molding varies not only in size but also in the angle that it makes with the wall. So the first step in calculating miter and bevel angles is to measure the crown with a framing square and determine the measurements shown in the drawing above. Then plug those figures into the formulas shown below.

What if you have a wall corner that is not 90°? To make this calculation you'll need a device for measuring the angle of the wall corner. I use the Angle Devisor (manufactured by Leichtung Workshops, 4944 Commerce Parkway, Cleveland, Ohio 44128; 800-321-6840). Whether you are installing inside corners or outside corners, be sure to use the angle of the inside corner (the angle less than 180°) for the equation.

Here's how the equation looks:

$$M = \tan^{-1}\left(\frac{A}{C \times \tan(F \div 2)}\right)$$

If we were to use our crown from drawing A, we would have 135 (F) divided by 2 = 67.5. Hit the tangent button and you get 2.4142. That times 4.8125 (C) = 11.6184. Divide 11.6184 into 2.875 (A), then hit the inverse-tanget button, and you get 13.9° (the miter angle).

For the bevel angle:

$$B = \sin^{-1}\left(\frac{D \times \cos(F \div 2)}{C}\right)$$

Plugging in some real numbers we get: 135 (F) divided by 2 = 67.5, the cosine of which is .3827. Multiply .3827 times 3.875 (D) and you get 1.4829. Divide that by 4.8125 (C), then hit the inverse sine button, and 17.9 appears. That's your bevel angle.

Finally, because the difference of one degree in the miter angle or bevel angle can be the difference between acceptable and unacceptable joints, you must set the angles on your compound-miter saw carefully. Math may be perfect, but measurements and the real world aren't, so slight adjustments may be needed to get an acceptable joint. But by using these equations you will avoid the fuss-and-fiddle approach I first used. □

Stephen Nuding is a carpenter in Hoboken, New Jersey. Photos by Susan Kahn.

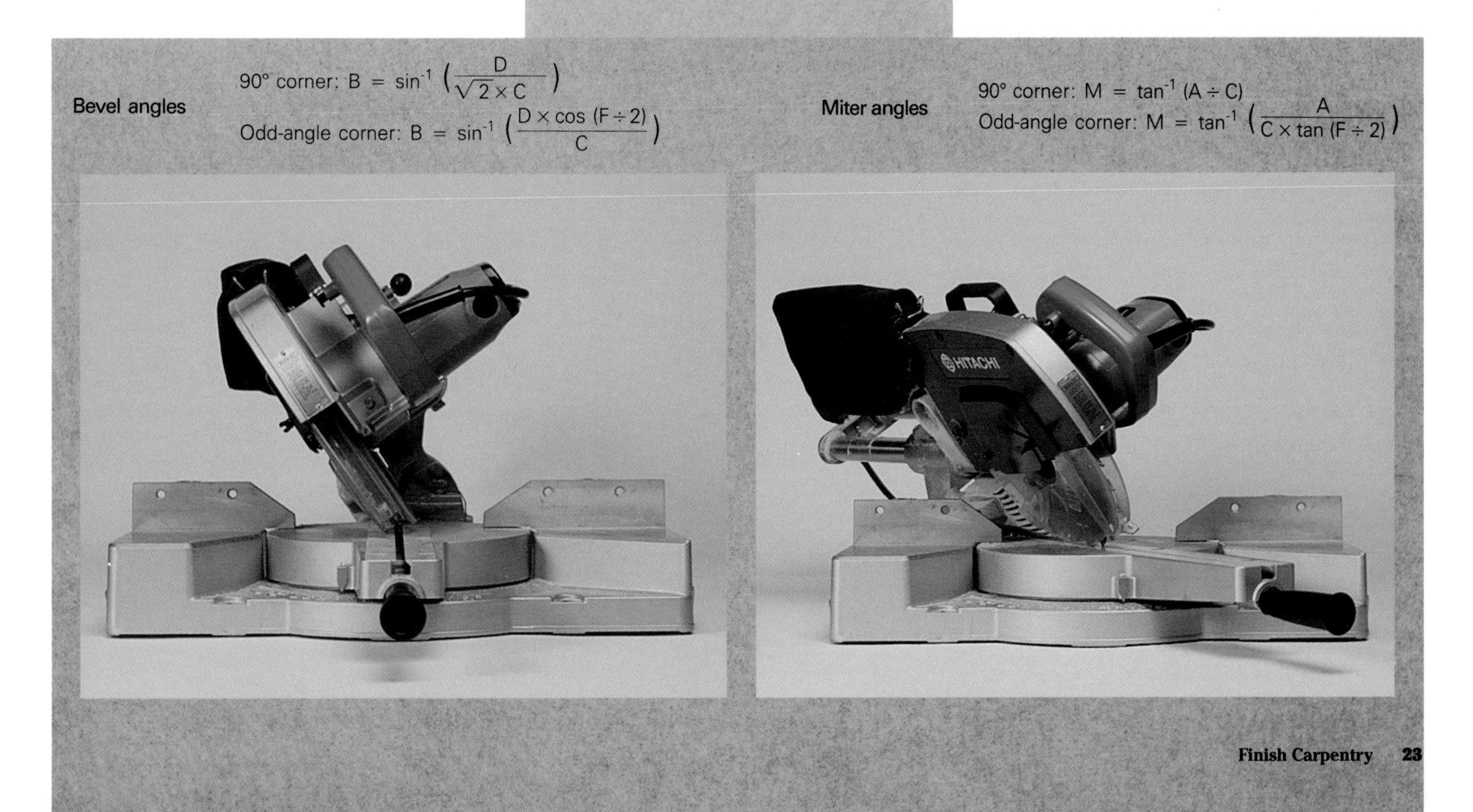

Hand Planes for Trim Carpentry

Tuned and adjusted right, these planes will save time and improve your work

by Scott Wynn

More than 100 different wood and metal hand planes are described in R. A. Salaman's book *Dictionary of Woodworking Tools: c. 1700-1970* (published by The Taunton Press, Inc.). Store-bought or handmade, many of these clever devices were once indispensable to builders. Before the advent of power planes and routers, a carpenter's repertoire might include assorted bench planes for preparing and smoothing wood stock; molding planes for shaping everything from stair nosings to door casings; and various contraptions for plowing dadoes, grooves and rabbets. A specialized carpenter might even own a compass plane for cutting convex or concave curves and a "galloping jack" plane for smoothing floorboards.

Nowadays, most of these planes are prized more by museum curators and tool collectors than by carpenters. But some types remain as vital on the job site as ever. As an architect/builder who specializes in trim carpentry, I use several kinds, primarily for fitting wood trim or casework against previously installed work, or wherever the use of a power plane or a router is impractical. My favorites are the block plane, the shoulder plane and the butt mortise plane. I also use an assortment of specialty planes (I made some myself) for cutting roundovers and chamfers.

Hand planes are available from woodworker's suppliers, mail-order tool outfits and some hardware stores and lumberyards. But don't expect planes to make smooth cuts straight out of the box. Properly tuned and adjusted, though, they'll cut wood like butter and sing while they work.

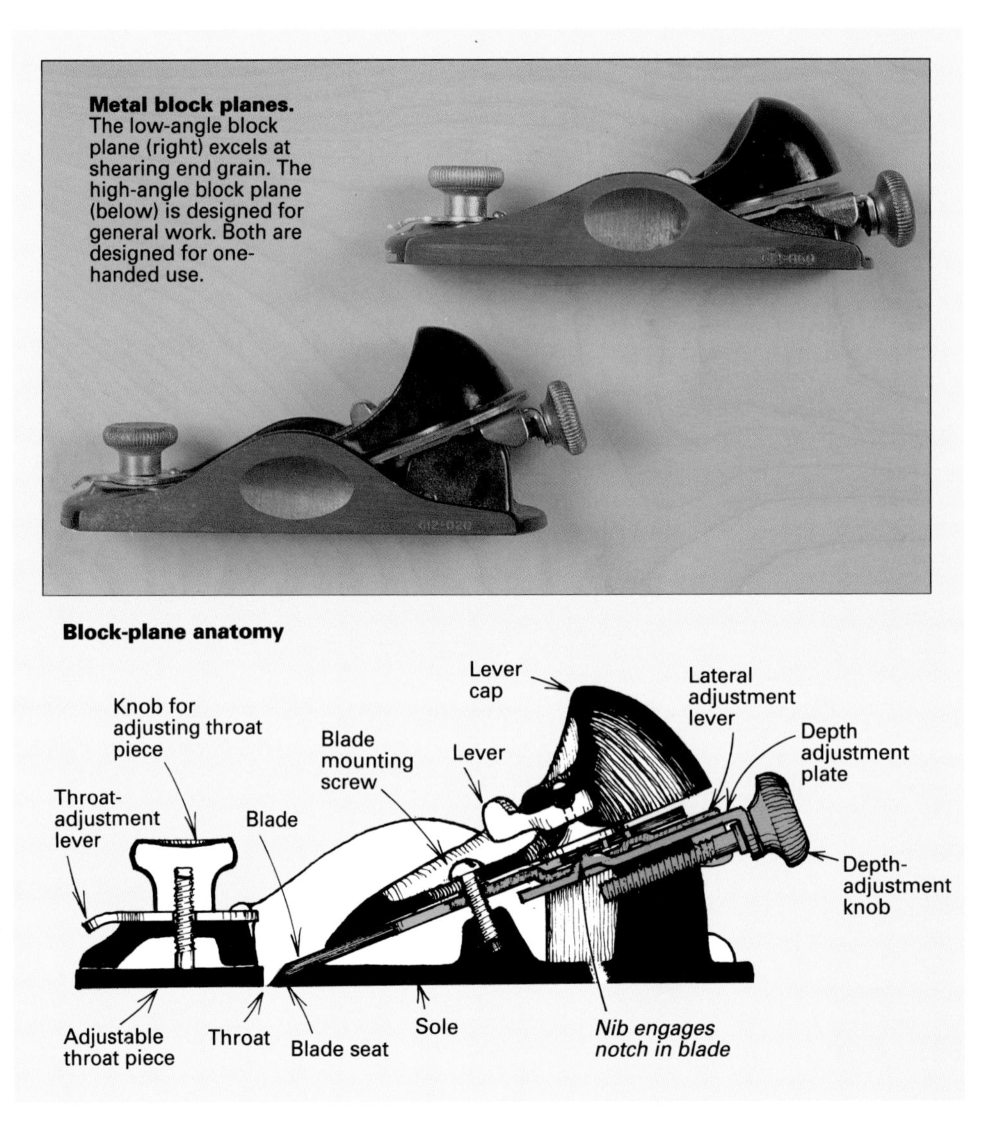

Metal block planes. The low-angle block plane (right) excels at shearing end grain. The high-angle block plane (below) is designed for general work. Both are designed for one-handed use.

The block plane—The typical metal block plane (drawing right) features an adjustable blade housed in a 6-in. to 7-in. long metal body. Mounted bevel-side up, the blade is clamped by a lever cap in two areas: against either one or two milled plateaus or a lateral adjustment lever at the top end of the blade, and against a narrow angled seat at the bottom end. The seat is directly behind the throat (the opening in the sole of the plane through which the blade projects). Depth of cut is controlled by turning a knurled nut or knob at the back of the plane.

Unlike its larger siblings—the jointer plane, the jack plane and the smoothing plane—the block plane is designed for one-handed use and will fit into most tool pouches. These attributes make it the plane of choice for most carpenters. I use mine for trimming miters, fine-tuning the fit of passage doors and flush-mounted cabinet doors, cleaning up jigsaw cuts, fitting cabinets to walls, flush-trimming screw plugs, planing door jambs flush with adjacent walls before installing casings and plenty of other routine tasks.

The planes that most carpenters are familiar with are the Stanley No. 12-020 and No. 12-060 (photo above) and the Record No. 09½ and No. 060½, though similar tools are made by other manufacturers (I own Stanleys). The 12-020 and the 09½ bed the blade at about 20°. The other two bed it at 12°. A low-angle plane is best for planing softwoods, hogging off wood and shearing end grain; a higher-angle plane cuts hardwoods with less tearout. Both types, however, perform so well when properly tuned that it's hard to tell the difference between them except under the most demanding circumstances.

All four of these planes have an adjustable throat, an important feature for preventing tearout—especially when making exceptionally fine cuts. Some block planes don't have an adjustable throat: Don't bother with them.

Body work—For a block plane to work right, its sole must be flat, its blade properly bedded, and the front edge of its throat must be smooth and parallel to the blade's cutting edge. The blade must be sharpened, with its back flat and free of imperfections (for more on sharpening, see the sidebar on p. 28).

From *Fine Homebuilding* magazine (August 1992) 76:80-85

Woodworkers have long debated the wisdom of flattening plane soles. Some argue that planes come flat enough from the factory, but I think a few minutes spent flattening a plane sole can improve performance significantly. Block-plane soles don't *really* have to be flat along their entire length. What matters most is that three areas of the sole contact a flat surface: the throat and both ends. If the throat area is relieved even slightly, the plane performance will be diminished.

Before flattening the sole, retract the blade but don't remove it. This way the plane body is stressed as it would be in use. I flatten the sole by rubbing it on a dry sheet of 600-grit wet-or-dry sandpaper laid on plate glass, being very careful not to rock the plane in the process. You can also use a saw table or a jointer bed instead of glass if you're sure they're flat (they usually aren't). The high areas of the sole will develop a dull, gray color that's easily distinguishable from the low spots. When the throat area and both ends of the sole turn this color, you're done. If at first the throat doesn't touch the sandpaper, I switch to 220-grit sandpaper to speed up the process, then to 320-grit, 400-grit and finally 600-grit paper once the throat makes contact. Finally, I smooth the edges of the sole with a file to remove any burrs or imperfections.

Next, inspect the blade seat to make sure that no burrs or bumps remain from incomplete milling. High spots can be leveled by removing the adjustable throat piece from the plane and flattening the bumps carefully with a fine file. If you don't see any bumps, don't touch the seat: You'll have a tough time restoring it if you mess it up.

Now mount the blade (and the throat piece if you've removed it) in the plane, sight down the sole and adjust the blade so that it protrudes $\frac{1}{32}$ in. or so, with the cutting edge parallel to the sole. Then adjust the throat piece so that it almost touches the cutting edge of the blade. Hold the plane up to a light and sight through the throat to make sure that the cutting edge is parallel to the edge of the throat piece. If it isn't, or if the edge of the throat piece isn't smooth and sharp-edged, remove the throat piece and file it where necessary. The throat edge must be straight and sharp. Do not round the edge of the throat piece, or the edge won't bear effectively on the workpiece to help prevent tearout.

Lastly, if you plan to use the plane with a miter-shooting board (see sidebar, p. 26), use a square to check that the sides of the plane body are relatively flat and are perpendicular to the sole. (If you plan to buy a new plane, check it for square in the store first so that you don't get stuck with a lemon). Carefully file off any high spots on both sides. Now the plane is ready for action.

Using the block plane—To use the block plane, mount the blade bevel up in the body and clamp down the lever cap, making sure that the lever cap's adjusting screw is tight enough to prevent the blade from being pushed around easily (but no tighter or you risk damaging the plane). Next, set the depth of cut and the throat opening according to the work you are doing. Flip the plane over, sight down the sole and adjust the plane so that the entire cutting edge appears at the throat as a black hairline. Now hold up the plane to a light source and adjust the throat piece so that the throat opening (the distance between the throat piece and the blade) is about $\frac{1}{32}$ in. for planing hardwoods or $\frac{3}{64}$ in. for planing softwoods. To combat tearout, the throat opening should be no wider than the thickness of the shaving. Test your settings by taking a few shavings from a wood scrap. For fine work, the shavings should be straight or rippled and thin enough to read through. If the throat jams, the opening is too narrow or the blade is set too deep. Adjust the plane and try again, repeating the process until you get the shavings you want.

The rabbet plane. **The Stanley No. 12-078 rabbet plane has two blade seats for regular rabbeting (above) or bullnose work. It comes with a cutting spur for cross-grain work, an adjustable fence and a depth gauge. Photo by Vivian Olson.**

The 3-in-1 plane. **The interchangeable nosepieces of the Record No. 311 "3-in-1" plane allow it to be used as a shoulder plane for rabbeting (below), a bullnose plane for working in confined spaces or a chisel plane for cutting stopped rabbets.**

The Japanese block plane. **Designed for maximum control, the author's Japanese block plane features a laminated-steel blade that holds an edge longer than western blades do.**

Making a miter-shooting board

Trimming small, short pieces of wood with a power miter saw is dangerous. A hand miter box won't trim less than a saw kerf's width (if that), and I'm not ready to buy a miter trimmer, a pricey tool that resembles a paper cutter. The solution to this dilemma is very old: the miter-shooting board, also known as a bench hook (drawings below). It's cheap, portable, safe and is less likely than expensive tools to walk away when your back is turned. Better yet, it can be used with a block plane, which lives in most carpenters' and cabinetmakers' tool kits.

I made my shooting board from scraps. The shooting edge should be made out of a durable material (such as ¼-in. tempered hardboard) glued to a ½-in. to ¾-in. thick plywood base. The miter block should be made out of a 1-in. thick composite material, such as particleboard or medium-density fiberboard (cross-grain movement of a solid-wood block would affect its accuracy). I cut the miter block using a power miter saw, then screw it to the base so that the block can be easily replaced if it's damaged or worn. A hardwood cleat is glued to the base so that the shooting board can be hooked over the edge of a worktable during use. I also rub a little candle wax to reduce friction where the plane will contact the board.

Before using the board, make sure your plane's sides are square to the sole. If not, file the sides until they are (mine only needed a touch-up in a few spots). Make sure the blade is sharp, and set it for a very fine cut with the throat open wide (tearout isn't a factor when planing across the grain). Then lay the plane on its side on the shooting board, making sure that it rests flat against the base of the jig and the shooting edge. Move the plane to engage the workpiece and then, with one firm stroke, remove a continuous shaving. Don't rock the plane during the stroke. Also, don't get a running start and crash into the piece, and don't chop at it. If you have to chop, either your blade is set too deep or it needs to be sharpened.

I usually take the board right to the area I'm working on so that I don't have to walk around after every stroke or two to check the fit. Also, with a little practice, you can tilt either end of the workpiece off the miter block to trim the piece for out-of-square conditions. *—S. W.*

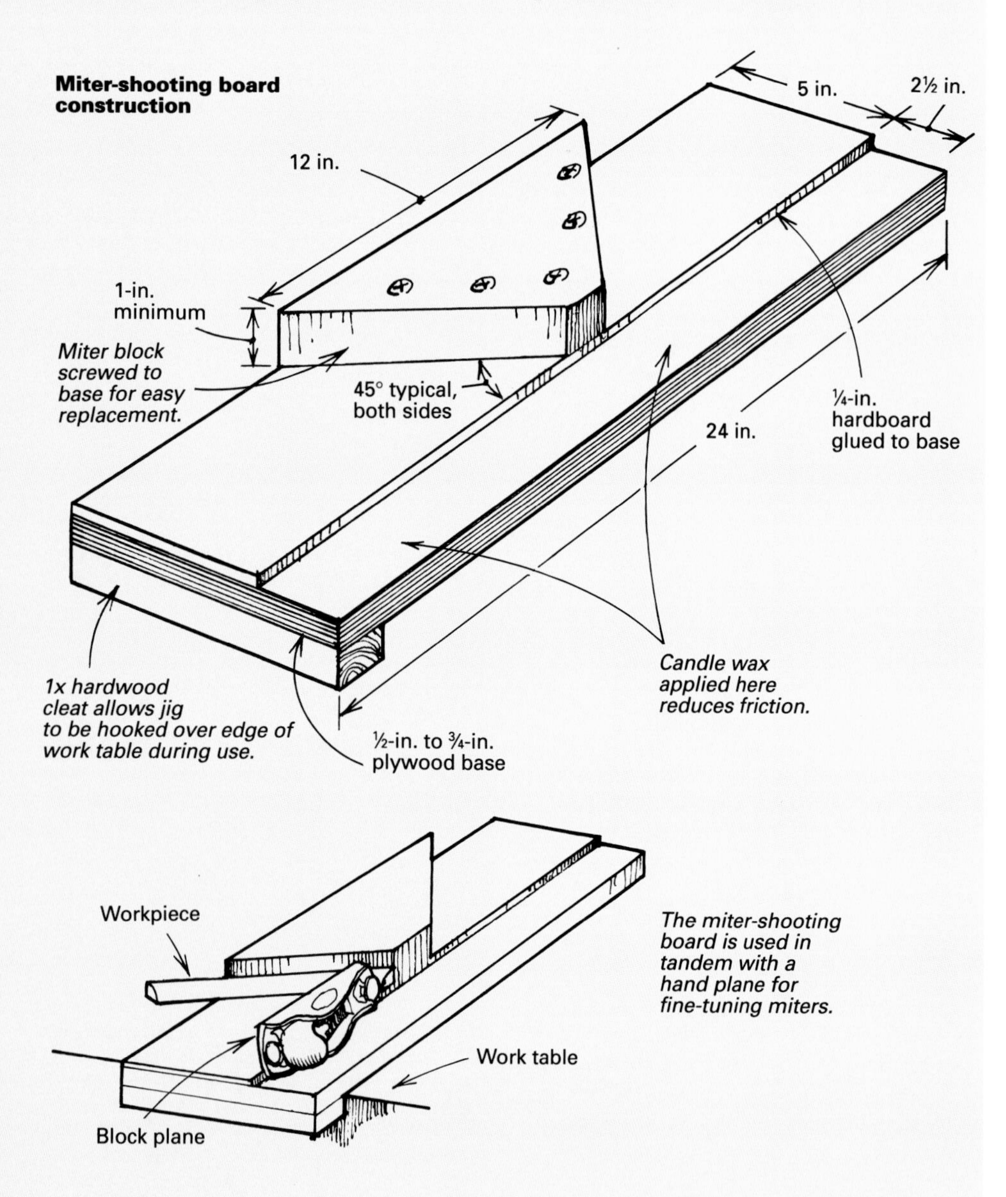

The miter-shooting board is used in tandem with a hand plane for fine-tuning miters.

Block planes will also hog off wood. To do this, open the throat about ⅛ in. wide to prevent overheating of the throat piece and the blade. Adjust the blade downward incremently until you get shavings of the desired thickness.

To preserve the cutting edge, don't bang it against the workpiece when beginning a cut, and don't drag the plane backwards along the work surface between strokes. Also, I always set down my planes on their sides, not on their soles.

The Japanese plane—Despite their versatility, my metal block planes have one drawback: limited durability of the cutting edge. Nowadays, there are high-quality aftermarket blades available that hold an edge longer than my stock blades do. One company, Hock Handmade Knives (16650 Mitchell Creek Dr., Fort Bragg, Calif. 95437), offers handmade, high-carbon-steel replacement blades for under $20.

Nevertheless, 16 years ago while searching for an alternative to my quick-dulling metal block planes, I bought a small Japanese plane (bottom photo, p. 25). Designed to be pulled instead of pushed, this plane has a 1¾-in. wide, laminated-steel blade wedged with a laminated-steel chipbreaker into a 7¼-in. long wood body (roughly the same length as my metal block planes). Though I often use my Stanley planes, I actually prefer using my Japanese plane on the job site. That's because it's lighter (I think of this every time I lift my toolbox), it fits comfortably into my small hands and my hip pocket, and it's surprisingly durable, having survived even a 35-ft. fall off a scaffold. Conversely, a short drop to a hard surface can crack an iron casting, usually at the throat, which renders the plane useless. I've also found that the plane's pull stroke gives me more control than the usual push stroke does (though I prefer pushing the plane when hogging off a lot of wood).

But the biggest reason I like the Japanese plane is edge durability. The secret to this durability is the marriage of a thin, extremely hard layer of high-carbon steel to a thick, strong layer of soft steel. The hard steel provides the cutting edge; the soft steel supports it. The cutting edge on my Japanese plane has actually shaved the very tops off nails (though the nail usually wins). I can use the plane all day, sharpen it that night and be ready for the next day.

Like their metal counterparts, Japanese planes must be tuned before use. The principles are similar—the blade must be sharpened and bedded properly, and the bottom must be flat—but the execution is a bit trickier. One excellent source of information on conditioning these planes is *Japanese Woodworking Tools: Their Tradition, Spirit and Use* by Toshio Odate (published by The Taunton Press, Inc.). Another wellspring of information on tuning and using hand planes, including Japanese ones, is *The Best of Fine Woodworking: Bench Tools* (also published by The Taunton Press, Inc.).

Japanese planes like mine cost about $45, comparable to the cost of metal block planes. They're available from a number of suppliers, including Hida Tool and Hardware Company, Inc. (1333 San Pablo Ave., Berkeley, Calif. 94702; 800-443-

Drawings: Scott Wynn

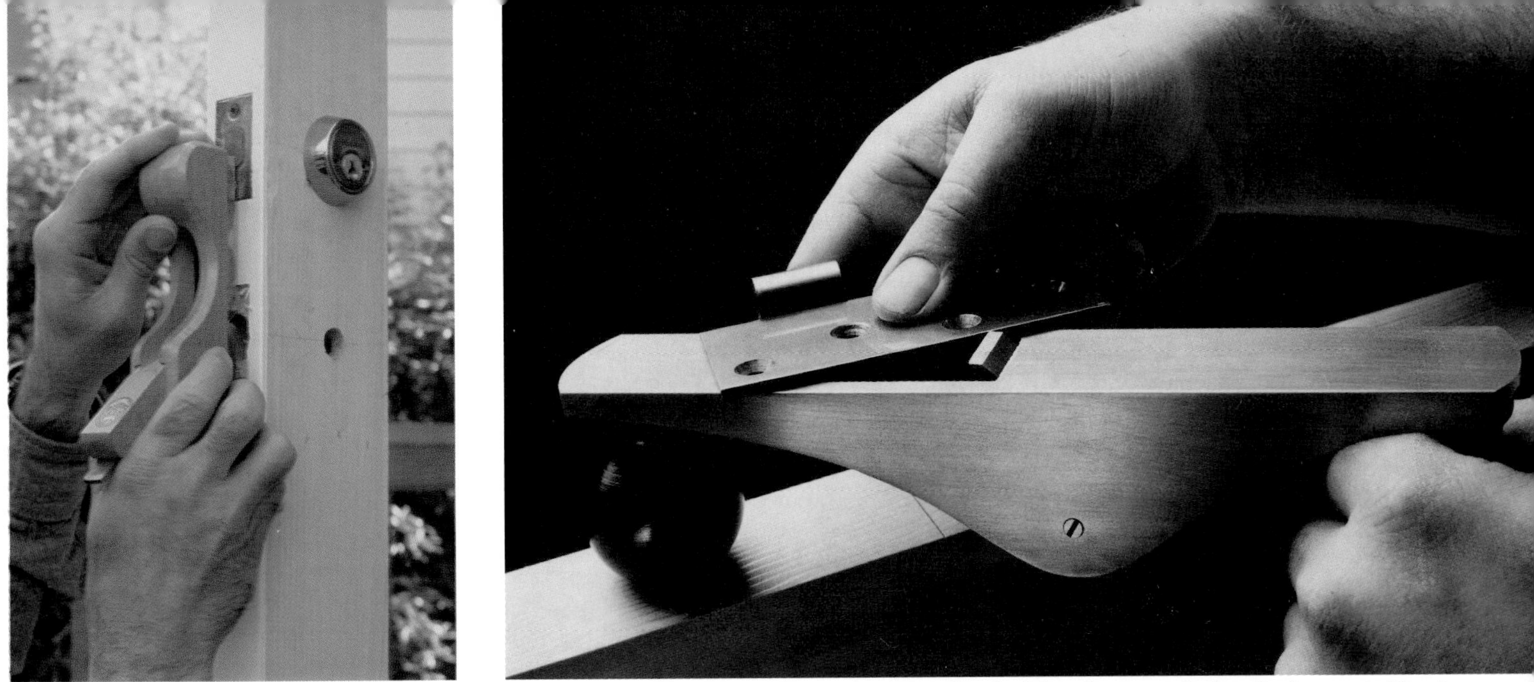

Butt mortise planes. **Author Scott Wynn's butt mortise plane (photo above left) cuts level mortises for hinge leaves and other flush-mounted hardware. Lie-Nielsen Toolworks makes the more common metal version (photo above right). Right photo courtesy of manufacturer.**

5512) and The Japan Woodworker (1731 Clement Ave., Alameda, Calif. 94501; 800-537-7820).

The shoulder plane—Block planes are the workhorses of trim carpentry, but a few other planes are worth having. I carry a shoulder plane for trimming rabbets because, although I rarely use it, sometimes nothing else will do. This is especially true when fitting new work to old. Rabbets are easily cut with a router or a table saw. But all too often new work is plumb or square, and the old work is not, so the rabbet needs a custom taper. This is easily accomplished using a shoulder plane.

My shoulder plane is an old Record No. 311 "3-in-1" plane (middle photo, p. 25). The 3-in-1 designation refers to three configurations accomplished through the use of interchangeable nosepieces. This allows me to install a long nose for shooting straight rabbets, a short nose for bullnose work in restricted areas, or to remove the nose altogether for chisel-planing to the end of stopped rabbets (rabbets that dead-end).

The 3-in-1 plane has become expensive since I bought mine and now costs about $150 (Clifton makes a similarly priced model). If I had to choose an alternative, I'd pick a Record No. 778 or a Stanley No. 12-078 rabbet plane (top photo, p. 25), which sells for about $65 to $75. Though it doesn't have a chisel-plane mode, it has a bullnose mode and a standard rabbeting mode. It also has a cross-grain cutting spur, a depth gauge and an adjustable fence, making it probably more versatile than the 3-in-1 plane. However, it's too large to fit easily into a toolbox, and it usually takes two hands to use, requiring the use of some clamping system to hold down the work.

Shoulder planes are generally machined more accurately than most brands of block planes, so unless you're having performance problems I wouldn't attempt to tune them. Trying to level the sole on a shoulder plane may tilt the sole out of square with the sides, which is a hassle to correct. Likewise, unless the blade obviously does not sit flat, I wouldn't touch the blade seat.

Chamfering and rounding over. **The author's collection of chamfer planes and rounding-over planes includes from left to right: a Radi Plane, which cuts roundovers having radii ranging from 1/16 in. to 1/4 in.; a 1/8-in. radius rounding-over plane; a 1-in. radius rounding-over plane; and an adjustable chamfer plane.**

With these types of planes, it's especially important to sharpen the edge of the blade square with the sides because there is virtually no allowance for lateral adjustment of the blades to compensate for an out-of-square cutting edge. The blade should protrude 1/64 in. or slightly less from either side of the plane body. Otherwise the plane will slowly step out from the shoulder of the rabbet as you plane.

The butt mortise plane—The butt mortise plane is used to cut level mortises for letting in hardware, such as hinge leaves, strike plates and dead bolts. Costing $45, the metal version made by Lie-Nielsen Toolworks, Inc. (Route 1, Warren, Maine 04864; 207-273-2520) resembles a standard plane except that it has a handle at each end, and its throat is wide open (top right photo). I own a rather obscure German wood model (top left photo) that I bought for $10 at a closeout sale. Like a metal plane, its long throat lets the chips pass through and allows you to watch what you're doing. Given the rather rough nature of mortising, there is no need to tune these planes beyond sharpening the blade.

When using my plane, I first lay out the mortise by outlining it with a chisel, then I make successive cuts with the chisel to the approximate depth required. At this point, the chips would normally be cleared out and the mortise leveled with the chisel. But I use the mortising plane. The blade depth is set to the thickness of the hardware (top right photo) and then the plane is pushed over the chisel cuts, popping out the chips much like a router plane. Then the plane blade is passed over the whole mortise again to remove any high spots. Lastly, the edges of the mortise are squared with the chisel.

The narrow body of this plane allows it to reach confined areas, such as mortises for hinges or strike plates in installed jambs with rabbeted stops. A router is certainly faster for production work, but if you have to cut a variety of mortises, hang a door in an existing opening or install a dead bolt in an existing door, the mortising plane will help speed things up.

Specialty planes—I carry other planes that can be time-savers (bottom photo, p. 27). I have an adjustable Japanese chamfer plane; a rounding-over tool called a Radi Plane (which cuts roundovers having radii ranging from 1/16 in. to 1/4 in.); a small Japanese-style, 1/8-in. radius, rounding-over plane that gives an exceptionally smooth finish; and a similar 1-in. radius rounding-over plane.

The chamfer plane allows me to match the chamfers that I machine in my shop on the edges of deck parts or trim, a boon if I need to produce an extra part on site. The Radi Plane and the small roundover plane duplicate the roundovers produced by some of my router bits, as well as those found on a variety of common moldings. The planes also allow me to match the slightly rounded edges typically found on flat stock. Like the chamfer plane, these planes allow me to avoid fussing with a router when I have to produce a simple edge detail or an extra piece of trim. The 1-in. roundover plane is pretty versatile in shaping a variety of radii that you might find on, say, stair nosings or door casings.

The Radi Plane costs about $22. Adjustable chamfer planes cost about $50. The rest are usually priced somewhere in between. □

Scott Wynn is an architect/contractor in San Francisco, Calif. He also designs and builds furniture. Photos and drawings by the author except where noted.

Sharpening plane blades

I know carpenters who hone their edge tools by rubbing them on two or three progressively finer sheets of soppy wet/dry sandpaper (ranging from 240 grit to 600 grit), taped or tacked to a scrap of plywood. I've heard of others who sharpen on their belt sanders. My sharpening system is more sophisticated than either of these methods, costs more and takes some time to master, but it produces a superb, long-lasting cutting edge that allows me to do top-of-the-line finish work.

Whatever sharpening system you use, I strongly discourage the use of honing guides. Feel the blade resting on its bevel and develop the body mechanics necessary to maintain that angle while sharpening. You may get frustrated at first, but you'll soon learn to get an adequate edge. As your woodworking skills improve, your sharpening skills will, too.

Sharpening stones—When it comes to producing a sharp, durable cutting edge with a minimum of effort, sharpening stones beat sandpaper every time. The selection of sharpening stones on the market is overwhelming. Oilstones have for generations been the mainstay in the West. Recently developed ceramic and diamond stones promise to combine many of the attributes of other types of stones with few of the drawbacks. For serious sharpening, though, I use Japanese water stones. Though they wear faster than other types of stones and must be flattened frequently, they cut very fast and produce an incomparable cutting edge.

Both synthetic and natural water stones are available. Synthetic water stones are less expensive and less fragile than natural water stones, but good-quality natural stones produce sharper and longer-lasting edges than the synthetic ones do. I use synthetic 1200-grit and 6000-grit water stones to sharpen American and European plane blades. For Japanese blades, I use the 1200-grit stone, an intermediate natural stone called an "Aoto Toishi" (or blue stone) and a deluxe 8000-grit synthetic finishing stone. On the job site, I use a diamond stone to touch up all my blades so that I don't have to deal with water.

Whatever stones you use, buy the best that you can afford. This is especially important for finishing stones, where price does equal quality. My water stones range in price from about $15 for the course stones to $50 for the fine ones. My fine-grit diamond stone cost $56. Most fine-woodworking suppliers carry a full line of sharpening equipment, including Japanese water stones. I got mine from The Japan Woodworker and Hida Tools (see addresses in text).

Grinders—Though you can get by without a grinder, if you use edge tools a lot you'll eventually want one. The two most common types are the bench grinder and the water-stone grinder. Bench grinders work okay, but you have to be very careful with them or they'll overheat the blade and draw its temper, destroying the blade's ability to hold a cutting edge. Bench grinders also hollow-grind the cutting edge, which leaves less metal on the blade than a flat bevel does for supporting the cutting edge. Japanese blades need a flat bevel because their hard, brittle steel at the edge requires the support of the softer, shock-absorbing steel laminated to it. However, any cutting edge subjected to hard use will benefit from a flat bevel.

The water-stone grinder (photo below) overcomes all of the shortcomings of the bench grinder. Water-cooled, it never overheats the blade. And because the revolving water stone is flat, you don't end up with a hollow grind. The water-stone grinder is ideal for beveling nicked or damaged blades. Its only drawback is that the rotating water stone wears out of flat quickly, requiring frequent trueing (I do this using my diamond stone). Water-stone grinders cost up to $300, but the ones I've seen will handle everything from chisels to planer knives.

The water-stone grinder. Water-stone grinders hone chipped blades quickly. An attached reservoir continuously dribbles water onto the rotating stone to eliminate the risk of overheating the blade.

The work area—Successful sharpening also depends on the nature of the tool-sharpening station itself. If you sharpen at a workbench, the stone should be 4 in. to 5 in. below your belly button. Unfortunately, the typical 3-ft. high bench is much too high for the average person. If the stone is too high, your wrist and elbows will be overly bent, and you'll have trouble maintaining a constant bevel on a blade. Also, your arms will do all the work without any help from your body weight.

I was taught to sharpen on the floor. Kneeling on a pad is pretty comfortable and often brings respite to a back tired from standing for long periods. The floor may be your only alternative on the job site, anyway. If you sharpen on the floor, elevate the stones about 6 in. Mine sit on a homemade redwood water trough (bottom photos, facing page), but you can also use a scrap of 6x6.

Whatever surface you work on, mount stops on it so that the stones don't move around during use, or use one of the manufactured systems that hold and store stones. You can keep synthetic water stones in a lidded plastic tub filled with water so that they'll be ready to go. Don't, however, store natural water stones in water or they'll disintegrate. They also may crack when frozen, even if they're dry.

Flattening the back—The first step in sharpening a new blade is to flatten and polish the back. Don't worry about polishing the entire back, however, just at minimum a narrow flat along the cutting edge (top left photo, facing page). I usually accomplish this using a steel lapping plate and silicon-carbide abrasive powders.

To use a lapping plate, pour ¼ teaspoon of 220-grit abrasive powder on the center of the plate and moisten the powder with a few drops of water (photo 1 below). Lay the blade backside down on the plate, perpendicular to the length of the steel, and rub the blade back and forth. Try to work all of the powder, including the piles that form at each end of the plate. The powder will eventually get very dry and fine, and the high spots on the back of the blade will start to get shiny (as opposed to the dull gray finish elsewhere). Continue rubbing until all of the silicon carbide is a fine paste (you may have to add a few drops of water now and then) and the back has a mirror polish along the entire cutting edge. Maintain even pressure at all times, and be careful not to lift the blade and round the edge. Backing up the blade with a stick helps (photo 2 below).

If the back of the blade is reasonably flat to begin with, I substitute a diamond stone, a 1200-grit water stone and a 6000-grit water stone for the lapping plate. The water stones must be dead flat, though. If a gray oval or large dot appears in the center of the stone while rubbing the blade on it, the stone needs flattening.

On the bevel—Now sharpen the bevel. Soak all but the finish water stones in advance until they stop bubbling (this takes just a few minutes). Sprinkle just enough water on the finish stones to create a slurry during sharpening.

To sharpen, grip the blade between the thumb and forefinger of the right hand (if you are right-handed), wrapping the other three fingers underneath the blade for support (photo 3 below). Holding the bevel flat on the 1200-grit water stone, press down on the edge of the blade with one or two fingers of the left hand and move the blade up and down the full length of the stone, gradually working from left to right and back as you stroke. Ideally, the cutting edge should be perpendicular to the length of the stone; in practice, it's easier to hold the edge diagonally. Keep the stone wet but not flooded. As you stroke, bend your arms and wrists to maintain the blade at the proper angle.

Check your progress by holding up the bevel to a light. The honed portion will be shinier than the unhoned portion. Also, check for a burr by brushing your finger away from the cutting edge. Once the bevel reflects light evenly (photo far right) and you can feel a burr along the entire width of the blade, move on to the blue stone (if you're using one) or to your finish stone. Don't exert as much pressure on these stones as you did on the 1200-grit stone; they polish more with the slurry formed than by direct contact with the stone. On the finish stone, back off (remove) the burr by laying the blade flat on the stone and rubbing it back and forth (photo 4 below). Then flip the blade over and polish the bevel. Alternate between the bevel and the back, shortening the number of strokes per turn until you finish with two or three light strokes on each side. There's no need to polish the edge further with a strop or a buffer.

A 30° bevel works best for most planing. This angle is easy to gauge: The length of the bevel is twice the thickness of the blade. If you're planing softwoods, a 25° bevel will cut cleaner and easier. Some people like to hone a secondary 5° microbevel on the cutting edge. I think this is self-defeating because the microbevel increases friction at the cutting edge and shortens its life. Besides, after the second or third sharpening, a microbevel becomes a macrobevel that requires nearly as much effort to sharpen as a full bevel.

Try to create a convex curve across the width of the blade while honing. This feathers the cut, eliminating steps or ridges across the surface. The curvature of the edge should be virtually indiscernable, equaling the thickness of the shaving you expect the plane to make. This way the blade will cut across its full width for maximum efficiency. An easy way to achieve curvature is by alternately applying pressure on one corner of the blade and then the other while sharpening.

While sharpening, check your water stones from time to time to make sure they're flat. One way to flatten them is to rub them on wet 220-grit or 320-grit wet-or-dry sandpaper laid on a flat surface (such as plate glass laid on a jointer table so that the glass doesn't flex). This technique tends to glaze the stones, however, reducing the cutting action until the top particles are worn away. I prefer to flatten my water stones with a diamond stone; it's quick and doesn't glaze the surface.

Before using the water-stone grinder, I saturate it with water. I don't use the bevel guide on the grinder. Instead, I simply feel the bevel, grinding perpendicular to the edge and moving the blade from side to side to wear the water stone evenly. Be careful that the grinder doesn't grab the blade and throw it, particularly when you first set the blade down. I don't use this grinder to flatten the backs of blades because it grinds too fast and may gouge the blade. *—S. W.*

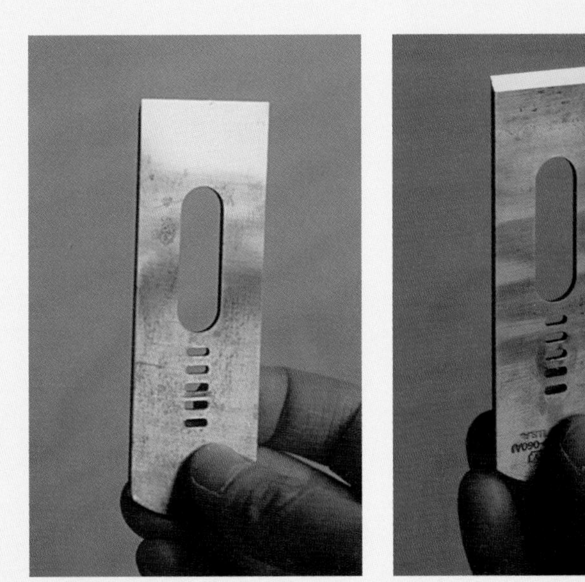

A sharp blade. The back of the well-tuned blade (above, left) is flat and polished along its entire cutting edge. The bevel (above, right) is honed to a mirror finish. Photos by Bruce Greenlaw.

1. Preparing a lapping plate.

2. Flattening the blade back.

3. Honing the bevel.

4. Backing off.

A Chop-Saw Workstation

A movable table for the compound-miter saw

by Scott King

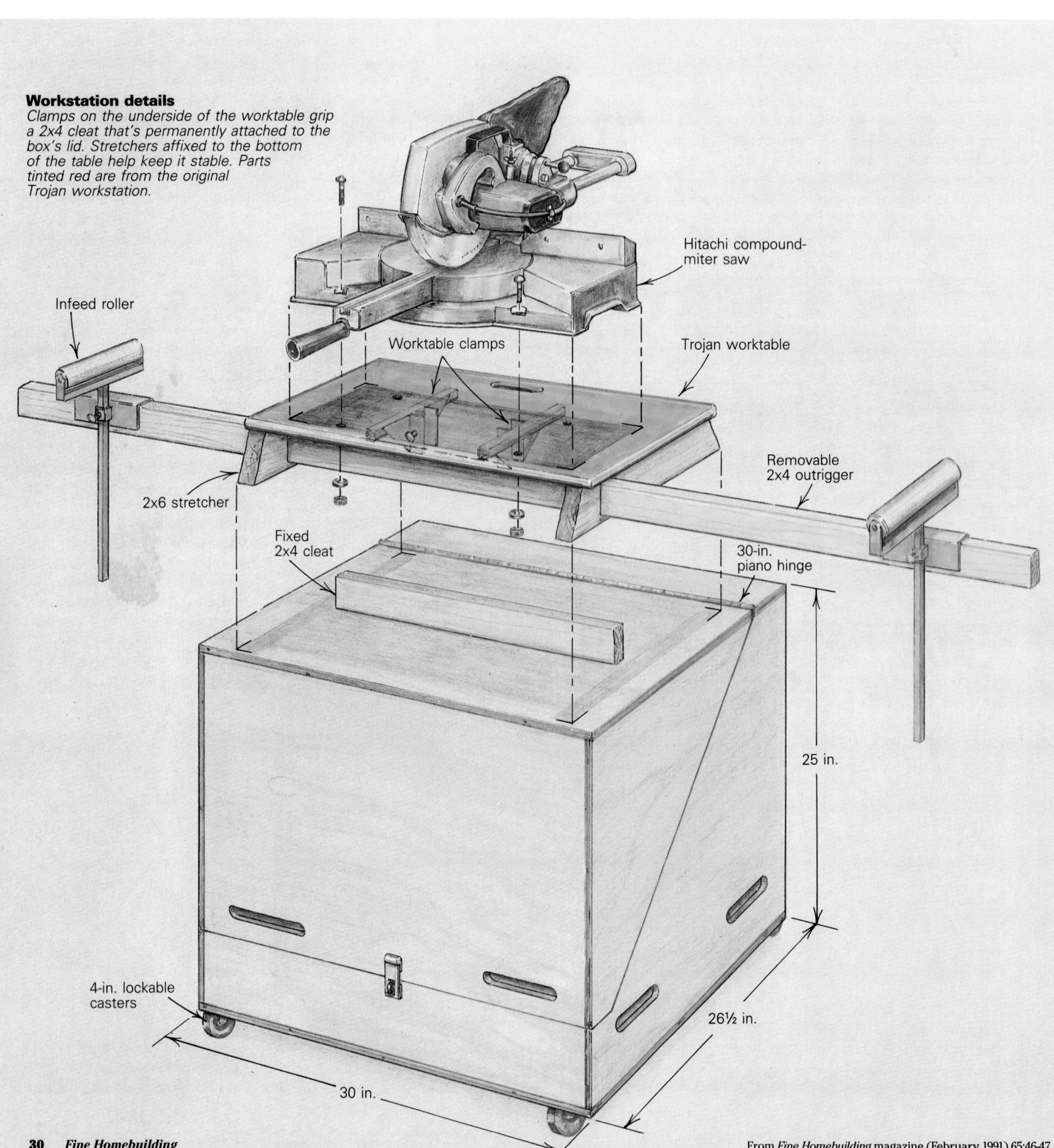

From *Fine Homebuilding* magazine (February 1991) 65:46-47

We've all been there. Nobody likes to use a chop saw on the floor, so the handiest thing around usually gets pressed into service for support duty. Sometimes it's a cabinet waiting to be installed; often, here in Bermuda, it's two barrels with a plank in between. Whatever you come up with, it usually doesn't move around much, so you end up bringing the work to it like a sacrifice to some ancient, immovable god.

When I bought Hitachi's compound-miter saw (see *FHB* #57, pp. 58-62), I was determined to break away from this mindless cycle and to build a portable work center (drawing facing page) that would take advantage of the saw's many features. As good as the saw is, in order to fully utilize it you'll need to bolt it down and provide some sort of infeed system to handle the long stuff. If you use the saw for long periods at a time, having it at a comfortable working height is also important. The ability to move the whole system around as work progresses and to provide secure storage when the saw isn't in use are the final criteria.

When I bought the saw I also bought a Trojan "Workcenter": a saw table and two adjustable rollers that clamp to the edge of a 2x6 and mount on a pair of metal sawhorse legs (Trojan Mfg., P. O. Box 15114, Portland, Ore. 97215; 503-285-2120). I thought it might work well with the Hitachi. I tried it out, but the Hitachi exerts quite a bit of leverage, and the table wobbled. For better results, I bolted a 2x6 base to the bottom of the Trojan table so it could sit on a flat surface. I then notched two holes in the base, just large enough to let a 2x4 pass through. The 2x4 acts as an outrigger for the adjustable-height rollers. Varying the length of the 2x4 or changing the placement of the rollers provides enough flexibility to handle most of the baseboard or molding we work with.

A movable beast. **As good as the Hitachi compound miter is, it's better when hooked up to a bench with an infeed system.**

The portable job boxes everyone has and some leftover ¾-in. plywood were all the inspiration I needed for the next step. I started by making a box 30 in. wide by 25 in. high by 26½ in. deep. Drywall screws and construction adhesive are all that hold it together, but after completing the box I belt-sanded the joints flush so it looked as if I spent a lot more time on assembly than I really did.

I cut the lid out of the completed and sanded box, remounted it with a 30-in. heavy-duty piano hinge, then added a heavy-duty hasp and a solid padlock to the front. The saw table is clamped to a 2x4 cleat that's permanently fixed to the lid (drawing facing page). If I need to get inside the box while the saw is on top, I just raise the lid, and the saw, table and outrigger assembly pivot up out of the way. Inside the box, two dividers create three bins for storage. When the saw is stored inside, these dividers raise it to the upper portion of the box, leaving ample room below for support rollers, extension cords and some additional tools (drawing left). The dividers also keep tools from shifting around during transit. I cut handholds in the sides of the box for lifting it in and out of the truck, and in the front for opening the lid. The latter is much appreciated when the saw and table are clamped to the top.

The box is mounted on heavy-duty, 4-in. lockable casters. The saw, table, base, box and casters reach a comfortable working height of about 35 in. I used movable casters on all four corners for maximum mobility and have never regretted it. Don't go cheap on this part because the ability to move the whole system smoothly around a tight job site littered with screws, scraps of wood and electrical cords really adds to its versatility and productivity. In hindsight, 5-in. or even 6-in. casters might have been worth the additional cost.

Bermuda has its own distinctive style of building, but when other carpenters or contractors see our rig, they are impressed with its convenience, mobility and versatility. Even the guys who have been doing this stuff longer than I've been alive admit that it would be nice to have "one of those fancy setups." And once they've used it, they never want to go back to a plank supported by two barrels. □

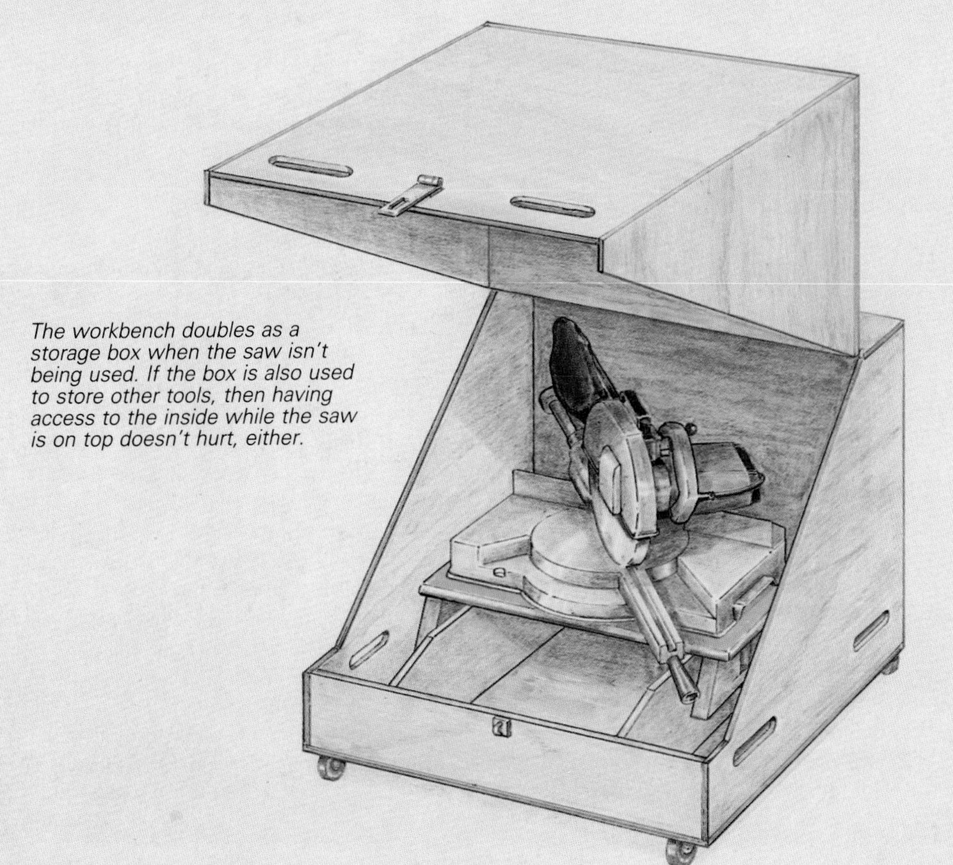

The workbench doubles as a storage box when the saw isn't being used. If the box is also used to store other tools, then having access to the inside while the saw is on top doesn't hurt, either.

Drawings: Bob Goodfellow

Scott King is a general contractor specializing in home renovation and commercial interiors. He lives in Bermuda. Photo by Victoria Epke.

Plunge Routers

Surveying the current crop, from the speedy flyweights to the rugged heavyweights

Routing made easy. **Unlike fixed routers, which have to be tilted into position for some cutting operations, plunge routers allow bits to be lowered into and retracted from a workpiece while the router base sits flat on the work surface. That's why plunge routers are superior for such tasks as cutting stopped flutes in door casings, as shown in the photo above.**

by Gregg Carlsen

Early in my carpentry career, I was confronted with the daunting task of building several flights of open-tread stairs. Each solid-oak stringer had to be mortised on site to accept a dozen 2x12 treads. Because the oak was dense, I routed each of the ¾-in. deep mortises in two passes. The days fell into a torturous monotony of positioning the mortising jig; clicking on the router; tilting the whirling dervish into place; routing the first recess; removing and turning off the router; lowering the bit a measured distance for the second, deeper cut; routing again; moving the jig; readjusting the router to the shallower depth of cut; and on and on.

Halfway through this ordeal, a cabinetmaker wandered over from the kitchen area and plunked a war-torn Makita plunge router in front of me. "Try this," he said. The tool won me over instantly. The router could be positioned, turned on, and then plunged straight down into the stringers slowly and surely. An adjustable pole-and-stop system allowed me to alternate between the two cutting depths with a quick click. "Hey, I like this!" Since then, plunge routers (specifically, a Hitachi TR-8 and a Makita 3612BR) have become essential parts of my power-tool arsenal.

Plunge versus fixed-base—Electric routers have been around since the 1920s, but it wasn't until 1949 that the German company Elu first devised a way to control the tool's downward plunge. At the moment, there are 30 models to choose from in the U. S., with power ratings ranging from 3.8 amps to 15 amps (see chart, facing page). All of them consist of a direct-drive, universal motor in a plastic and metal housing that rides up and down a pair of spring-loaded, tubular posts affixed to a sturdy base. This configuration, as opposed to that of fixed-base routers, allows plunge routers to be positioned flat-footed on a workpiece with the router bit fully retracted, and then plunged straight down, compressing the post springs in the process. During the routing operation, the motor can be locked at the desired depth of cut by tightening a locking lever or a rotating handle (top photo, p. 34). When the operation is complete, the lock is released and the motor is allowed to scale the base posts, lifting the router bit clear of the workpiece.

All plunge routers come equipped with an adjustable metal pole that works in concert with a stop or a series of stops (in the form of a multi-step rotating turret) to limit the depth of cut (bottom photo, p. 34). This system allows multiple passes at various depths with a minimum of hassle. On most of the routers, upward travel of the motor is adjustable and is limited by either a large twist knob or a pair of lock nuts on a threaded rod. In some cases, the lock nuts can also be used for firmly locking the router into a fixed-base configuration.

Plunge routers not only put less strain on the operator and reduce the likelihood of nicking a workpiece; they also minimize the hazard of accidental startups. In fact, for safety, parts of Europe now mandate that *all* routers have a convenient system for automatically retracting router bits from the workpiece.

The utility of plunge routers is readily apparent in the cabinet shop, where dadoes, mortise-and-tenon joints and template work are a way of life. But plunge routers are also becoming increasingly popular for both rough and finish carpentry. I've found them particularly useful for milling trim (photo above), da-

Photo by Nanny Carlsen.

Plunge-Router Specifications

Model number	List[1] Price	No-load RPM	Motor (amps/HP)	Weight (lb.)	Collet capacity (in.)	Spindle lock	Plunge depth (in.)	Base shape	Accessories included
AEG OFS 50	$299	25,000	6.25/1	5.5	¼	yes	1 15/16	RF2	A, B
Elu 3303	$235	24,000	6.5/1	6	¼	no	1 15/16	RF2	A, B, C
Elu 3304[3]	$292	8,000-24,000	5-6.5/1	6.2	¼	no	1 15/16	RF2	A, B, C
Elu 3337	$379	20,000	12/2.25	11.25	½	yes	2 7/16	RF1	A, B, C
Elu 3338[3]	$427	8,000-20,000	10-12/2.25	11.25	½	yes	2 7/16	RF1	A, B, C
Bosch 1611	$386	22,000	14/3	12	½	yes	3	RF1	B, C
Bosch 1611EVS[3]	$448	12,000-18,000	15/3.25	12.25	½	yes	3	RF1	B, C
Freud FT 2000	$350	22,000	15/3.25	12.5	½	yes	2¾	RF1	B, G
Hitachi TR-8	$215	24,000	6.9/1.25	6.4	¼	no	2	RF2	A, B, C, D, I
Hitachi M 8	$215	25,000	7.3/1.5	6.4	¼	yes	2	RF1	B
Hitachi M 8V[3]	$278	10,000-25,000	7.3/1.5	6.6	¼	yes	2	RF1	B
Hitachi TR-12	$347	22,000	12.2/3	11	½	no	2⅜	R	A, B, C, D, E, G, H, I
Hitachi M 12SA	$403	22,000	14.6/3	11.5	½	yes	2 7/16	RF1	A, B, C, D, G, I
Hitachi M 12V[3]	$447	8,000-20,000	15/3.25	11.7	½	yes	2 7/16	RF1	A, B, C, D, G, I
Makita 3620	$192	24,000	7.8/1.25	5.7	¼	no	1⅜	RF2	B, J
Makita 3612B	$340	23,000	14/3	12.7	½	yes	2½	S	B, G
Makita 3612BR	$340	23,000	14/3	12.5	½	yes	2½	R	B, G
Metabo 0528	$279	27,000	5/.75	7.75	¼	yes	2	RF2	A, B
Porter-Cable 693[2]	$295	23,000	10/1.5	11.5	½ and ¼	no	2½	R	B, F
Porter-Cable 7538	$410	21,000	15/3.25	17.25	½	no	3	R	B, F
Porter-Cable 7539[3]	$475	10,000-21,000	15/3.25	17.25	½	no	3	R	B, F
Ryobi R50	$177	29,000	3.8/.75	5.1	¼	no	2¼	RF2	A, B
Ryobi R150K	$201	24,000	6.5/1	6.2	¼	no	2	RF2	A, B, J
Ryobi R500/R501[4]	$331	22,000	13.3/2.25	11	½	no	2⅜	RF2	A, B, D, E, G, H
Ryobi R600	$398	22,000	15/3	13.6	½	yes	2⅜	R	A, B, D, G, H
Ryobi RE600[3]	$444	10,000-22,000	15/3	13.6	½	yes	2⅜	R	A, B, D, G, H
Skil 1823	$98	25,000	8.5/1.5	7	¼	yes	2	RF1	B
Skil 1835	$99	25,000	9/1.75	7	¼	yes	2	RF1	B, C
Skil 1870	$245	23,000	12/2.25	9.5	½	yes	2½	RF1	B, C

Notes:

1. Retail price is typically 30%-50% lower.
2. Model 6931 plunge base can be purchased separately.
3. Features electronic speed control.
4. Identical models, except for switch location.

Base shapes:
R = Round
S = Square
RF1 = Round with one straight edge.
RF2 = Round with two straight edges.

Accessories:
A. Straight-edge fence
B. Collet wrench(es)
C. Universal template adaptor
D. Template guide
E. Roller guide
F. ¼-in. collet
G. ¼-in. collet adaptor
H. ⅜-in. collet adaptor
I. Carbide bit
J. Tool case

doing stair stringers and plunge-cutting rough openings for doors and windows in wall sheathing. In fact, plunge routers can handle any task that a fixed-base router can.

Plunge routers do have a few drawbacks. Because their handles are secured to the motor housing rather than to the base, they force your hands to ride several inches above the work surface when the router isn't fully plunged, creating a high and somewhat unstable center of gravity. Also, due to their up-and-down plunging action, these routers tend to vibrate more than fixed-base routers do, making their multitude of lock nuts and clamp screws prone to loosening (that is, if they aren't tightened properly). That's why regular inspections are a must. Finally, plunge routers typically cost about 20% to 30% more than comparably sized fixed-base routers.

Sizing them up—Shortly after I agreed to evaluate plunge routers for *Fine Homebuilding*, I received one each of every professional and serious DIY-grade model currently available in the U. S. (except for a few that virtually duplicate sister models, plus or minus a feature or two). For a tool-junkie like me, this was the next best thing to Christmas (never mind that the routers weren't mine to keep).

The sheer number and diversity of plunge routers on the market dictated that I find a convenient way to group them for ease of comparison. Though manufacturers typically market their routers according to horsepower, I decided to group them by amperage because amps are a more reliable gauge of a motor's performance. I divided the routers into three categories: under 8.5 amps; 8.5 to 13.3 amps; and 14 amps and over. Keep in mind that these groupings are for convenience only. Based on overall performance, some of these routers are clearly a notch above or below their peers; I'll point out these tools along the way.

With the help of two co-workers who specialize in custom carpentry, I edge-molded and plunge-cut oak and pine using a variety of carbide router bits. To test the firmness of the various settings, we turned on the routers and plunged and retracted them repeatedly in a seat-of-the-pants torture test. Finally, we field-tested the group of routers for three weeks to see what they're really made of.

Scanning the features—There were no real dogs when it came to motors. From the $98 Skil 1823 to the $475 Porter-Cable 7539, *all* of the routers cut smoothly. Within each amperage category, however, the tools differed considerably in concept and overall performance.

During our evaluation, we sized up all the basic features: plunging mechanisms (posts and springs), plunge-limiting devices (adjustable poles and stops), plunge locks (levers and rotating handles) and height-limiting mechanisms (knobs and lock nuts). But there's more to the story. For instance, most of the on/off switches on these routers are easy to reach without letting go of a router handle. The ones that aren't are a bit awkward to operate, making them potentially dangerous.

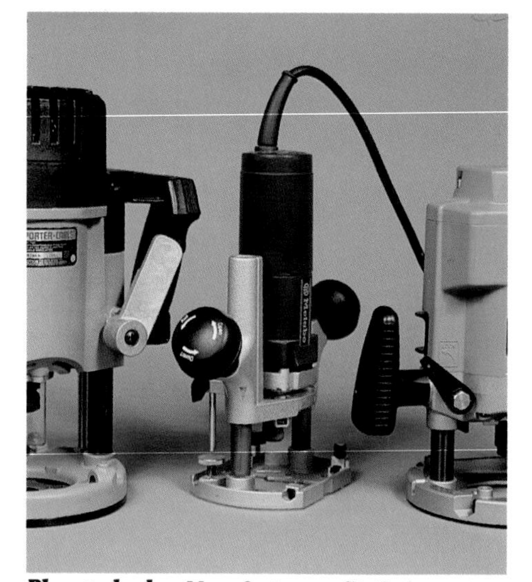

Plunge locks. **Manufacturers fit their routers with one of three different types of plunge-locking mechanisms: (from left to right in photo above) a spring-loaded lever that unlocks when it's depressed; a handle that tightens when it's rotated clockwise 1/16 of a turn; or a spring-loaded lever that locks when depressed.**

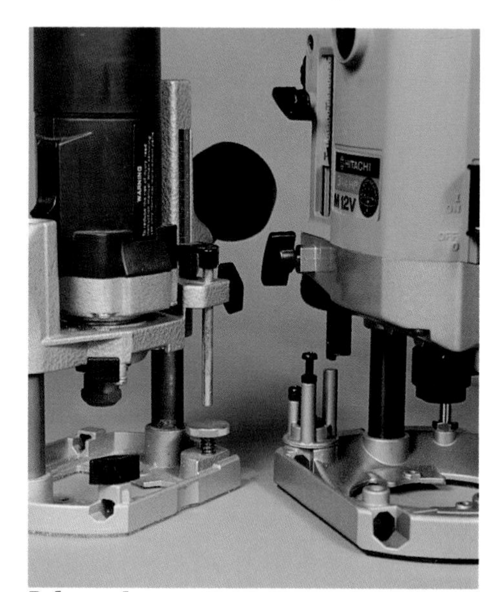

Poles and stops. **Without exception, depth of cut is controlled by an adjustable metal pole that works in concert with a single stop (left in photo) or by a multi-step rotating turret (right in photo) affixed to the router base.**

Base shapes vary from round to square, with a couple of compromises in between. Round bases are superior for guiding routers against irregularly shaped templates, while square ones are more compact and are best for working against a straight edge. Round bases having one or two straight edges are a sensible compromise.

Weight is a tradeoff; light routers are more maneuverable than heavy routers, but heavy routers are more stable. Collet sizes also vary; generally, the more powerful the router, the bigger the router bits it will accept. Most carpenters I know have a ¼-in. router bit collection that far exceeds their ½-in. collection. However, all ½-in. plunge routers accept not only ½-in. shank router bits, but (when fitted with accessory collets or collet adaptors) ¼-in. and sometimes even ⅜-in. bits as well. Spindle locks are another handy feature. They allow router-bit installation using one wrench instead of two. That's one less wrench that can slip and damage the plunge posts, and one less to keep track of. Eighteen of the routers we evaluated have spindle locks.

Several plunge routers come equipped with variable speed, electronic speed control and a "soft-start" feature. These features add about 10% to 30% to the list price of a router, but I think they're worth it. Variable speed allows you to work at the most efficient rpm for most routing chores. This often translates to lower rpm and less noise, not to mention less wear and tear on the router. Electronic speed control maintains the selected speed for optimal cutting, regardless of load. The soft-start feature gently accelerates router motors to full speed, particularly nice with the high-powered production models.

Finally, there's cost. As usual, you get what you pay for. When comparing list prices on the chart, don't ignore the ""Accessories included" column. The Hitachi TR-12, for instance, comes with virtually everything but a tool case (the straight-edge fence alone is worth about $20, discount price), while the Hitachi M 8 comes with a collet wrench only. Also, keep in mind that mail-order prices are typically 30% to 50% less than list prices.

Under 8.5 amps

The high rpm and small bases of the routers in this group (photo facing page) make these tools particularly well suited for laminate trimming, decorative plunge-cutting and light edge work (such as rounding over and chamfering)—especially in tight quarters. These little routers have oblong bases (round with two parallel straight edges) except for Hitachi's new M-series routers, which have round bases with one straight edge. Both base shapes conform equally well to straight or curved templates. All the routers in this group hold ¼-in. shank router bits only, and make it quite clear by chattering and screaming when they're being pushed to the limit.

AEG OFS 50—This machine is impeccable. Its on/off switch and plunge-lock lever operate smoothly and are accessible from the handles. A spindle lock eases router-bit installation.

Plunge depth is controlled with a three-step rotating turret and a metal pole that's raised and lowered by turning a plastic rotary dial on the side of the motor housing. A depth scale molded into the dial reads in millimeters and 64ths of an inch.

The standard fence included with this router is substantial, and includes a superb fine-tuning mechanism for setting the guide a precise distance from the bit. The router also accepts a flexible shaft that holds a variety of carving and grinding bits. AEG will special-order the shaft on request (AEG Power Tool Corp.; 800-243-0876. In Connecticut, 203-447-4600).

The flyweights. **Left to right, top row: Metabo 0528; Hitachi TR-8; Hitachi M 8V. Middle row: Ryobi RI50K; Ryobi R50. Bottom row: Elu 3304; Makita 3620; AEG OFS 50.**

Elu 3303 and 3304—Though Elu has been selling plunge routers in Europe for more than 40 years, it wasn't until 1987 that Black & Decker finally introduced Elu routers to the U. S.

Elu's models 3303 and 3304 are almost identical except that the former is a single-speed unit, while the latter sports variable speed, electronic speed control and soft start. The variable speed ranges from 8,000 rpm to 24,000 rpm—the widest speed range available on the market. You can really hear the 3304 accelerate to maintain constant rpm through knots and twisted grain.

Depth of cut on both routers is controlled by means of a multi-step turret and a sliding pole that points to a depth scale. The pole is locked in place by a simple clamp screw. This system is a bit touchy when it comes to fine-tuning, but it does the job.

The on/off switches are easily controlled with the left hand. The right-hand handles double as plunge locks that require about 1/16 of a full turn to activate, a feature I quickly learned to like. Changing bits requires the use of two wrenches; a spindle lock would really give these routers a boost (Black & Decker U.S. Power Tools; 800-762-6672).

Hitachi TR-8, M 8 and M 8V—The TR-8 is the old standby of Hitachi's small-router line, and the M 8 and M 8V are recent upgrades. The M 8 is a single-speed unit, while the M 8V features variable speed, soft-start and electronic speed control. Hitachi intends to continue offering all three models.

The M-series routers are not only more powerful than the TR-8 but they also offer a number of features that the old model lacks, including spindle locks, adjustable handles, an easily accessible on/off switch, and a nifty new adjustable pole that works in concert with a three-step turret for setting the plunge depth. Like AEG's depth pole, this one employs a rotary knob for raising and lowering it. When pulled out about ½ in., however, this same knob raises and lowers a separate zero-reference depth scale. This system allows you to adjust the depth of cut by plunging the router until the bit contacts the workpiece, lowering the pole until it contacts a stop on the turret, setting the depth scale to zero, and then raising the pole so that it points to the desired depth-of-cut reading on the scale. The system is accurate and easy to master.

At first I thought Hitachi's new handles, which adjust to three different angles, were a gimmick. But once I adjusted the handles full tilt and started working at floor level I grew to appreciate their comfort. Changing the angle requires the use of a Phillips screwdriver and takes about 10 seconds.

All three of these routers have a threaded column with lock nuts on it for setting maximum height of the motor assembly or locking the machine into a fixed-base router, a feature not found on most of the other small routers.

My main complaint with these routers is that they don't plunge smoothly unless the base is fully supported, which can be troublesome when edge-routing. This isn't necessarily a safety problem because edge-routing can be accomplished with the router base locked in the fixed position, but it's a nuisance nonetheless. Both M-series routers also had weak springs—neither router returned to full height consistently (Hitachi Power Tools U. S. A., Ltd.; 800-829-4752).

Makita 3620—This router cuts exceptionally smoothly. Still, it seems to be designed more for the weekend woodworker than for the serious amateur or pro. For instance, the 3620 is the only plunge router with a trigger-type on-/off switch, which has to be gripped constantly to activate the motor. Also, the router's molded uni-body construction may eliminate a lot of screws that could rattle loose, but it also means you have to replace the entire body if the handle or any other part breaks. The router comes with a molded plastic case that would be more useful if it included a few slots for storing router bits.

The 3620 has a shallower maximum depth of cut than any other plunge router on the market. My biggest complaint, however, is the tool's uneven plunging action when the side of the router opposite the plunge lock is unsupported. Even with the base fully supported, I had to favor the plunge-lock side to achieve a smooth plunging action (Makita U. S. A. Inc.; 213-926-8775).

Metabo 0528—The model 0528 is Metabo's sole offering in the router field, but the company does plan to make bigger models.

I had a love-hate relationship with this router. On the positive side, the 0528 is locked in

The middleweights. **Clockwise, from top left: Hitachi TR-12; Elu 3338; Porter-Cable 693; Skil 1870; Skil 1823; Ryobi R501.**

a plunged or retracted position by turning the right-hand handle about 1/16th of a turn. It comes with a well-designed dust-collection attachment that can be connected to a wet-dry vacuum (AEG, Bosch, Elu, Hitachi and Skil offer accessory dust-collection attachments). The depth-of-cut system includes a sliding pole (held fast by a clamp screw) and a single adjustable stop that still allows you to make roughly measured multiple passes before making a final accurate cut. The router's high rpm makes it particularly useful as a laminate trimmer.

But the router has several annoying features. For instance, to turn it on, you have to push the switch down and pull it out using the thumbnail of your left hand. My router also had lazy post springs. Before I'd shell out the money for this router, I'd want to see the switch and spring action improved (Metabo Corp.; 215-436-5900).

Ryobi R50 and R150K—Looking a little like the robot R2-D2 from Star Wars, the R50 is the lightest-weight plunge router on the market, has the least powerful motor and spins at the highest rpm (a whopping 29,000). Despite it's small size, the router is smooth and spunky. The motor can be removed and inserted into Ryobi's fixed base (R30) or laminate trimmer base (R70), making it one of the most versatile routers in the group. The plunge springs are exposed. This produces no significant negative effects, except that the springs tend to jamb up with sawdust and gunk more often than internal springs would; however, they are easy enough to clean.

The R150K has more power than the R50 and, though it doesn't have uni-body construction, is similar to Makita's 3620. It's a good basic machine that's geared primarily to the weekend woodworker (Ryobi America Corp.; 800-525-2579).

Between 8.5 and 13.3 amps

This category of routers (photo above) encompasses a wide range of performance. The Elu and Porter-Cable mid-size entries are bonafide workhorses, for example, while Skil's 1823 and 1835 perform more like lighter-duty tools. In fact, Skil's two light-duty entries also have just 1/4-in. collets, while the rest of these routers have 1/2-in. collets.

Elu 3337 and 3338—These two routers are virtually identical except for their electronics. The 3337 is a single-speed unit, while the 3338 features variable speed, soft-start and electronic speed control, all of which performed flawlessly.

Both models have the look and feel of a high-quality tool. The rotary wheel used for raising and lowering the depth-stop pole operates smoothly and precisely. The zero-reference vernier scale, which comes complete with a built-in magnifying glass, is a beauty. Height adjustment is controlled by a large knurled nut that spins up and down a threaded rod. A quick-release button on the nut disengages the threads, allowing rapid adjustments.

Both routers are equipped with a spindle lock to simplify bit installation and a collet nut with a special "double clutch" grip that virtually eliminates router-bit slippage. When removing a bit, the collet loosens, then tightens again after a three-quarter turn, then loosens again and releases the bit. Other routers seem to offer this feature, too (though in many cases it's tough to tell), but Elu is the only manufacturer that explains it in their manual.

Hitachi TR-12—Though Hitachi rates the TR-12 at 3 hp, it draws just 12.2 amps, placing it squarely in the mid-range category (all the other 3-hp plunge routers on the market draw between 14 and 15 amps). This router's motor operates smoothly, but the plunge action is tentative and uneven, especially when the base is only partially supported. The TR-12 is a good basic machine, but if I were bent on purchasing a large Hitachi router, I'd plunk down the extra $50 to $100 and step up to the newer, more powerful M-series routers (more on those later).

Porter-Cable 693—This router combines Porter-Cable's 10-amp model 6902 router motor with its model 6931 plunge base. The base, which can be purchased separately for less than $100 (list price is $120), is fitted with an aluminum carriage that accommodates most 3½-in. dia. Porter-Cable (or Rockwell) motors from their line of fixed-base routers. It also holds the motors from Bosch's 1600-series routers, Black & Decker's 2720 router and a few Sears units.

The rugged 6931 base incorporates a spring-loaded plunge-lock lever that automatically de-

The heavyweights. **Clockwise, from top left: Porter-Cable 7539; Hitachi M 12V; Bosch 1611EVS; Ryobi RE600; Makita 3612B; Freud FT 2000.**

faults to the locked position and must be pressed down to allow plunging, a nice safety feature. Depth of cut is controlled by a standard pole-and-stop system that features a zero-reference scale and a rotating turret having three fixed and three adjustable stops. The carriage plunges with just the right amount of spring action, whether or not the base is fully supported. The height of the motor is limited by means of two knurled nuts that ride on a threaded rod at the top of the right-hand plunge post.

The only feature that's even moderately flaky about the marriage of this base and motor is the location of the on/off switch, which is near the top of the motor housing. However, even this switch can be operated from the router handle by a long, strong left thumb (Porter-Cable Professional Power Tools, Inc.; 901-668-8600).

Ryobi R500 and R501—The only difference between these two models is the location of the on/off switch: on the R500 it's mounted on the motor housing, while on the R501 it's on the handle. These routers are simple, no-frills machines that perform smoothly and efficiently. Drawing 13.3 amps, their motors are just a step below heavy-duty routers. Plunge return is limited by a large twist knob that's particularly useful for adjusting depth of cut when the router is mounted in a router table. The routers come loaded with accessories: a straightedge fence, roller guides, template guides and $\frac{1}{4}$-in. and $\frac{3}{8}$-in. bit adaptors.

The Ryobi R500 that I tried had one serious flaw. Though fresh out of the box, the machine wouldn't retract after plunging until I tapped on both sides of the motor housing with a screwdriver. The folks at Ryobi told me the problem was uncharacteristic of their routers and that I had received a bad unit.

Skil 1823, 1835 and 1870—According to Skil, the 1823 and 1835 routers are designed for serious do-it-yourselfers. In other words, don't plan on using them for serious production work. Nevertheless, they do have some nifty features: both have a built-in compartment for stowing the collet wrench, and the 1835 has room in the handle for storing three router bits. Both routers have external springs and a single stop block molded into the base. For the part-time user, $100 (list price) buys a tool that will handle those occasional odds and ends.

The Skil 1870 is the freshest face in the plunge-router crowd. Its single-speed motor runs smoothly and delivers adequate power. The molded handles are substantial and comfortable, offering easy access to the plunge-lock lever and to a handle-mounted trigger switch. The unit I evaluated had one serious flaw, however—during rigorous use, I repeatedly knocked the plunge pole out of adjustment. Given that the pole has a very smooth finish and is secured with an inconvenient little lock knob, this didn't surprise me. Also, unlike those on all the other plunge routers, the power cord on the 1870 exits straight out the right-hand side of the motor housing. During routing, this cord kept getting in my way.

My router *was* one of the first 1870s off the assembly line (in fact, it has "#33" imprinted on it), so chances are, refinements are forthcoming. According to Skil, a variable-speed version will be available soon (Skil Corporation; 312-286-7330, product department).

14 amps and over

All six brands of routers in this category (photo above) are designed for production work. Their rugged, powerful motors will handle most cutting chores without bogging down or overheating, even during continuous use. Their mass helps to dampen vibration, contributing to smooth cutting action.

Weighing anywhere from $11\frac{1}{2}$ lb. to $17\frac{1}{4}$ lb., these routers can wear you out when working with them on vertical surfaces or overhead. But when using them for flat work, they're less tiresome to use than the smaller routers because they cut faster and with less vibration. I'm especially fond of the soft-start capability of some of these machines because it eliminates having to wrestle with the gyroscopic action of their huge motors. These routers also plunge to depths ranging from $2\frac{3}{8}$ in. to 3 in. This capacity is a must for using some of the longer router bits on the market.

Bosch 1611 and 1611EVS—These two models appear to be almost identical, except that the EVS wears a 1-in. high cap that houses the electronic brains for variable speed, soft-start and speed control. The EVS has an extra $\frac{1}{4}$ hp, too, primarily to boost the speed control when it's pushed to the limit.

I asked the folks at Bosch why they designed the EVS to run between 12,000 and 18,000 rpm—relatively slow speeds for a multiple-speed router. They said that most people use big variable-speed routers for powering hefty router bits, such as raised-panel cutters, and that routers used for this purpose perform better and last longer when operated at lower rpm (after all, most shapers run between 7,000 and 10,000 rpm). The single-speed 1611 runs at 22,000 rpm.

Like Porter-Cable, Bosch uses spring-loaded plunge-lock levers that lock when you let them go, handle-mounted on/off switches and a threaded rod with stop nuts to limit return height. I don't like Bosch's nine-step turret; there are so many stops on it that I was afraid I might turn to the wrong one in a production situation. The turret also doesn't click securely at each stop like the turrets do on the other plunge routers.

These routers are the only plunge routers that are available in 220-volt models for shop use. According to Bosch, these routers will soon be replaced by two new models—the 1615 and the 1615 EVS—that will resemble their predecessors, but feature improved bearing systems, smoother plunging action, and other amenities (Robert Bosch Power Tool Corp.; 919-636-4200).

Freud FT 2000—Freud has been in the tool business for just a few years, but the Spanish company that manufacturers their power tools (which Freud now owns) has been in business for 100 years. The FT 2000, Freud's sole plunge router, was introduced to the U. S. market about two years ago.

This router's 15-amp motor is a solid performer, and its plunge action is first rate. I especially like the big twist knob on top that limits the upward travel of the motor. It's particularly handy for setting the depth of cut when the router is mounted in a router table. The depth-stop consists of a threaded pole that contacts a three-step turret affixed to the base. The pole is captured by a nut that doubles as a spring-loaded, quick-release button. Pressing the button disengages the threads, allowing rapid adjustment of the pole up or down. Fine-tuning is accomplished by simply turning the pole.

Unfortunately, other features were a headache to use. When inserting the ¼-in. collet adapter, there is nothing to prevent it from disappearing way up into the router's ½-in. collet. Even when I inserted the router bit into the adapter before inserting the adapter into the collet, as the people at Freud suggested, I still had trouble. The separate little warning card included with the router indicates that I'm not the first to complain.

The plunge-lock lever is locked by lifting it, a movement that I found awkward and susceptible to accidental releases (by bumping it loose). Also, during some routing operations, one of the stops on the rotating turret can actually depress the quick-release button on the plunge-depth pole—an irritation to say the least (Freud, Inc.; 919-434-3171).

Hitachi M 12SA and M 12V—The extra 2-in. high motor cap on the M 12V houses electronics for variable speed, speed control and soft-start, features the M 12SA lacks. Otherwise these two routers are identical and offer the same features as Hitachi's smaller M 8 and M 8V routers.

The plunge action of these heavyweights is the best of the Hitachi routers, but could still use a little improvement with the base only partially supported. Both routers come with all sorts of goodies, including a sturdy adjustable fence, a template guide, a ¼-in. collet adapter and a ½-in. carbide straight bit.

Makita 3612B and 3612BR—The sole difference between these two models is the shape of their bases: the BR model has a round base, and the B model has a square one. At 23,000 rpm, the 3612 operates at the highest rpm in its class. The motor is smooth and powerful; during one test I inadvertently plowed, with no chatter or strain, a ½-in. wide by 1¼-in. deep groove through a Hemlock fir plank. The router also plunges and retracts smoothly, except when only the right-hand side of the router is supported.

Smooth and strong, or fast and maneuverable?

The plunge-lock lever and on/off switch are relatively small, but easy to operate. The depth-stop is a threaded pole with a quick release button, same as the Freud.

Porter-Cable 7538 and 7539—The two "Speedmatic" routers are part of a new generation of plunge routers introduced by Porter-Cable in the past two years. Both models feature larger armature shafts and heftier bearings than their predecessors for increased reliability and reduced vibration. They also feature plenty of metal; at 17¼ lb., the Speedmatics are the heaviest plunge routers on the market.

Though the 7538 and 7539 appear almost identical, the former is a single-speed machine surmounted by a circuit breaker, while the latter is a five-speed unit with a speed-control dial on top. The 7539 also features soft-start and electronic speed control.

The Speedmatics have the same type of spring-loaded plunge-lock levers and the same type of pole-and-stop system for controlling depth of cut as Porter-Cable's model 6931 plunge base. A threaded rod with stop nuts limits the plunge return.

The plunge posts are hardened and protected by a special heat-treating process. The posts slide through Teflon-lined bushings in the motor housing, producing a fluid plunging action. These routers had the snappiest return action of all; I nearly gave myself a sex change while testing the plunge-lock lever on one of them. The on/off trigger switches are mounted on the handles and feature lock-on buttons that allow the option of working in controlled bursts or continuously.

I've owned a Porter-Cable 3-hp, fixed-base router for years, run it for hours on end and cut miles of ¾-in. plywood; it's never so much as sneezed. The new Speedmatics are smoother yet, and are clearly no-nonsense machines.

Ryobi R600 and RE600—Ryobi offers two versions of their top-of-the-line machine. The R600 is a single-speed unit, while the RE600 features variable speed, soft-start and electronic speed control. Working at 16,000 rpm (sufficient for many heavy-duty woodworking operations), the RE600 hums like a high-quality vacuum cleaner (mine was the quietest of the routers I evaluated), a nice alternative to being screamed at all day. Ryobi's three-step turret and simple depth-adjustment pole are as basic as they come, but held up firmly in the torture test. Plunge return is limited by a twist knob near the top of the router. The spindle lock is a simple, unobtrusive button.

One complaint: to plunge smoothly, I had to favor the right-hand handle slightly. Plunging with the right-hand side unsupported was especially awkward.

Ryobi is in the process of developing an even heavier-duty plunge router—a 3½ hp bruiser that should hit the market soon.

Personal favorites—Any purchasing decision will ultimately boil down to the power you need and the features you like. Some routers are heavily geared toward production, while others are geared more to lighter-duty use. The big routers run smoother and stronger; the small ones run faster and are more maneuverable. You'll have to decide which is right for you.

I loved working with the Elu 3338. It's relatively lightweight, has plenty of power, features variable speed, handles both ¼-in. and ½-in. shank bits, is easy to adjust and is relatively quiet. I knew I was handling a well-designed machine.

For heavy-duty production work, I like Porter-Cable's Speedmatic models. My older model 520 fixed-base Speedmatic has served me well for years, and these plungeable versions are better yet.

If you don't anticipate heavy cutting but are striving for accuracy, consider the little AEG OFS 50. Its easy-to-tune depth adjustments and ability to take a flexible shaft make it ideal for detailed, light work.

Finally, I'd recommend Porter-Cable's 6931 base to anyone who already has a compatible motor in his fixed-base router. It's an ingeniously designed accessory that's priced right. □

Gregg Carlsen is a builder and writer who lives near St. Paul, Minnesota. Photos by Susan Kahn except where noted.

Cranking Out Casements

Using a layout rod and shaper joinery to build outward-swinging windows with divided-light sashes and insulated glass

by Scott McBride

One of the loveliest houses in Irvington, New York, was built in the 1920s by a wealthy philanthropist named Ralph H. Mathiessen. The house commands a breathtaking site on the banks of the Hudson River, overlooking the Tappan Zee. It's an elegant example of the French eclectic style, characterized by decorative brickwork and a steep, slate-covered roof.

Unfortunately, by the time the present owners acquired the house, successive waves of "modernization" had spoiled much of the original detailing. Among other things, casement windows in the third-floor dormers had been traded for aluminum jalousies. I was asked to undo some of the damage by replacing the replacements with true divided-light windows. In addition, the owners wanted to upgrade the thermal efficiency as much as possible, so new glazing had to be insulated glass. None of the new windows we needed were available in stock sizes, and while some of the major window manufacturers now do custom work, the need to save time and money prompted me to build them myself. Over the course of the project, I fabricated sidelights and French doors as well as casement windows using the same basic joinery. But I'll limit the discussion here to casements.

Casement hardware—The job was a first for me, so I familiarized myself with the anatomy of casement windows by reading up on the subject in A. B. Emary's *Handbook of Carpentry and Joinery* (Sterling Publishing Co., out of print) and Antony Talbot's *Handbook of Doormaking, Windowmaking, and Staircasing* (Sterling Publishing Co. Inc., 2 Park Ave., New York, N. Y. 10016, 1980. $8.95). Then I turned to the hardware catalogs to see what kinds of fittings were available.

Casement windows can either swing in or out. Outward-swinging casements (photo right), which I built for this job, are easier to make watertight because rain driven between the sash and the frame drains out rather than in. They are also more convenient to use because they don't project into the room when open, as do inward-swinging casements.

The chief disadvantage of outward-swinging casements is a susceptibility to decay. If left open, even in a light rain, the sash will be soaked. Another problem is that the screen must be placed inside the window, blocking access to the sash. At one time screens were hinged so that you could get at the old sliding-rod hardware used to open and close the windows. Modern casements open and close by means of a worm-gear crank mounted on the stool. Only occasional access to the sash is required, so the screens can be snapped in place.

Sliding-rod and crank-type hardware (known as "casement operators") is still available from H. B. Ives (a Harrow Co., P. O. Box 1887, New Haven, Conn. 06508). The crank-type operator is by far the more practical, so I incorporated it into my design for the Mathiessen house windows. Although we wanted brass to match the rest of the hardware, Ives makes the operators only in painted steel.

Modern casements have a lever-type lock that operates with the screen in place, but my customers wanted the look of traditional brass casement locks (called "fasteners"), which were available from Ives. Since the worm gear of a crank-type casement prevents the window from being forced open, even when unlocked, the main purpose of the fastener is to provide a tight weather seal.

Casement frames—Fortunately, many of the original cypress window frames in the dormers were reusable. Most of these had simply been boxed around to make a finished opening for the aluminum jalousies. Where the frames were damaged beyond repair, I removed them and made new ones.

The new side and head jambs have two rabbets each—one on the outside for the sash

Built in the 1920s, this house suffered subsequent remodeling that saw the original casement windows in the dormers replaced by aluminum jalousies. But the wheel of architectural fashion has now come full circle and the casements are back; this time with double-pane insulated glass.

From *Fine Homebuilding* magazine (December 1988) 50:69-73

and one on the inside for the screen. These rabbets could have been cut into thick stock, but I found it easier and more economical to attach a ½-in. thick stop to ¾-in. thick jambs.

For the window sills, I decided to use cypress. Although some folks say the cypress available nowadays has grown too fast to contain the resin needed for decay resistance, I gave cypress the benefit of the doubt. The good condition of the original casement frames argued strongly in its favor. I cut a pair of shallow rabbets in the sills, known as "double sinking," to help to keep rain from blowing under the sash (drawing below). A drip groove on the bottom of the sill prevents water from being drawn under the sill by capillary action.

Designing for insulated glass—By far the trickiest part of the job was designing and building the new sash. Switching from ⅛-in. single-pane to ½-in. double-pane insulated glass required an overhaul of the original sash design. Unlike single-pane glass, which is sealed with a thin bead of putty, insulated-glass units are typically held in place by moldings because the solvents in putty will attack the rubber seal around insulated glass. Consequently, a rabbet for insulated glass must be deep enough to accommodate both the increased thickness of the glazing and the thickness of a wood molding. Although the original casement sashes in the Mathiessen house were a hefty 2¼ in. thick, the rabbet for glass was only ¾ in. deep—not enough to hold insulated glass. For this reason I made replacement sashes for some of the remaining original wood casement windows, as well as for those that had been switched to aluminum.

At my local supplier, I had a choice of either 8/4 or 12/4 rough stock. I decided to build the new sash from 8/4 stock planed to 1¾ in. On the outside of the sash, I cut a ⅝-in. deep cove-and-bead profile on a shaper. This profile is called *sticking* because it is worked directly on the sash—"stuck" rather than applied. The glass rabbet took up the remaining 1⅛-in. thickness of sash: ½ in. for the glass and ⅝ in. left for an applied molding similar in profile to the cove-and-bead sticking.

Muntins make up the inner framework, or grid, that holds the glazing in a divided-light sash. On this project, the width and depth of the muntins had to be carefully considered. I was doubling the weight of the glass, so stronger muntins would be needed. In addition, the applied molding had to be wide enough to hide the ½-in. deep rubber seal that separates the two panes in an insulated-glass unit. But if the muntins were too wide, the sash would look more like a dungeon grate than the delicate tracery I wanted. I set-

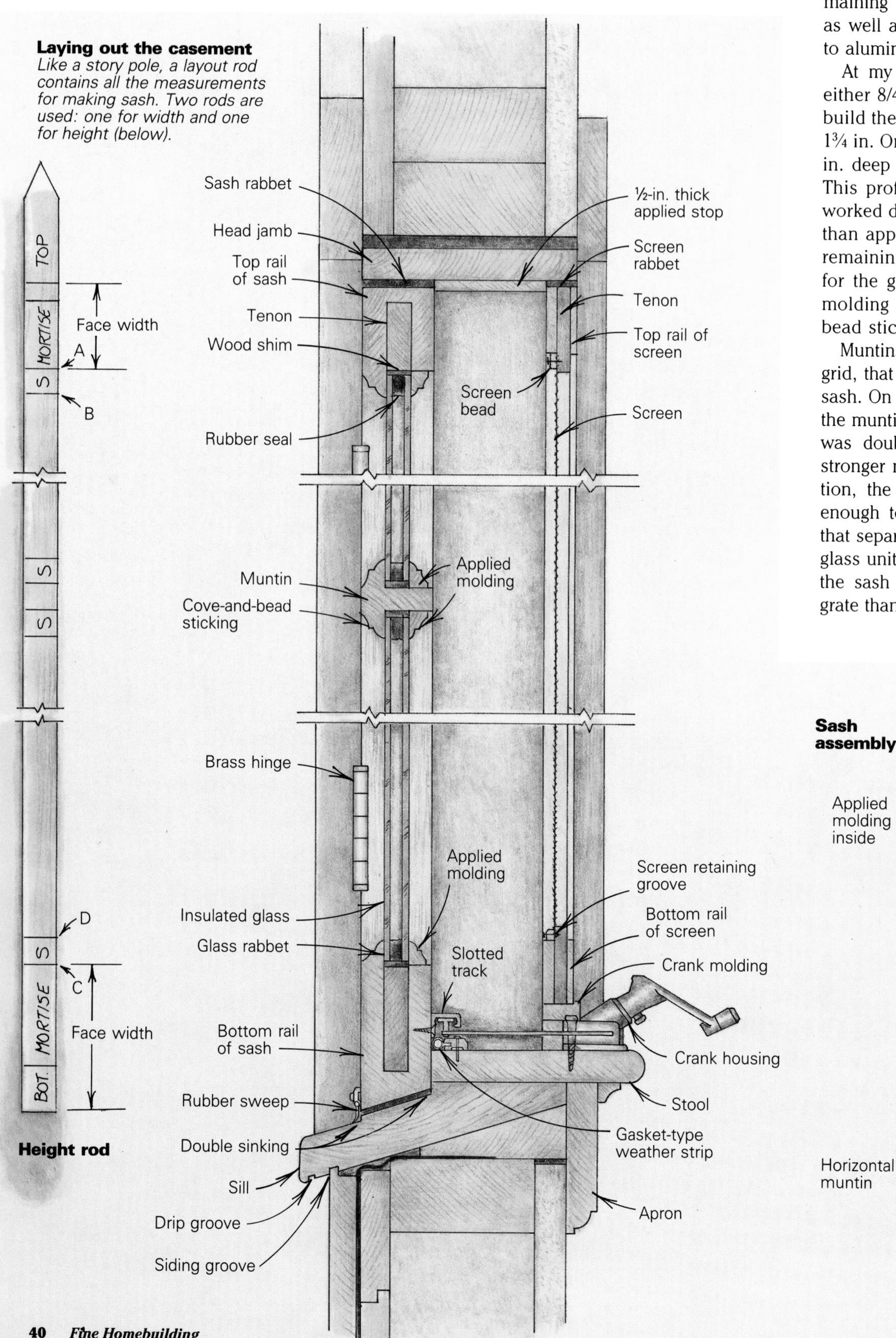

tled on a muntin width of 1½ in. This would accommodate a ½-in. wide flat face down the middle, with ½ in. sticking on either side.

Choosing materials—I chose white pine for the sashes, jambs and trim. This old standby is stable and easily worked, and is fairly decay resistant. Sash work requires a lot of short pieces, so I figured I could cut around the knots in Grade 2 material without too much waste. I saved the long, vertical-grained pieces from the edge of each board for stiles and used the less stable flat-grain stock from the middle of the board for the rails and muntins. All this selective cutting took time though, and I've since found it cheaper to buy clear sugar pine in the first place.

I got my insulated glass from a local glazier, who gets it from a wholesale fabricator. Quality control was less than rigorous, however, and I returned a lot of defective units. The defects included corners out of square and ruptures in the rubber seal. I was particularly careful to root out the latter—I've no desire to be replacing fogged-up units a year or two down the road.

Rod layout—Like a story pole for a house, a rod is a layout stick containing all of the pertinent measurements for making a sash. You need two rods—one for height and one for width—and laying them out is the single most important step in sash-making. It requires the builder to think through the entire cutting and assembly process before any sawdust flies.

I began the layout of the height rod by cutting a piece of 1x2 a few inches longer than the height of the sash. I cut one end square and marked it with an arrow and the word "bottom." Starting from this end, I laid off the overall height of the sash, marking this line with an arrow and the word "top."

From the top, I measured back a distance equal to the face width of the top rail, made a mark (point A, drawing facing page), then added the sticking width and made a second mark (point B). I repeated this for the bottom rail (marking points C and D).

I measured the distance from C to A and deducted from this measurement the combined face width of all horizontal muntins. For the sash in this example, the combined thickness is 1½ in. (3 x ½ in.). I divided the balance by the number of lights the sash has along its length (four). The result was the distance from rail to muntin and muntin to muntin, measured on the inside of the sash. This could also be expressed as the distance from shoulder of glass rabbet to shoulder of glass rabbet. After determining the measurments, I marked them on the rod. Then I added the sticking width (½ in.) to both sides of each muntin, marking it with an "S."

Because hinged casement windows are subjected to a racking force resulting from their own weight, they are best made with mortise-and-tenon joints, rather than with the bridle joint sometimes used for double-hung sash. So the next step in preparing the height rod was to lay out the height of the mortises to be cut in the stiles. Although I wanted maximum surface area in the joint for gluing strength, I was careful to leave enough wood (½ in.) between the mortises and the ends of the stile to prevent the wood from fracturing.

I labeled the mortise heights on both ends of the rod. The smaller mortises—those for the stub tenons in the horizontal muntins—had essentially been laid out already because their height is the same as that of the muntin face width, which I had marked earlier.

When I'm building larger casements or divided-light doors, I usually have a through-tenoned horizontal muntin in the center, tying the stiles together and making the sash more rigid. I didn't think that was necessary here. The casement sashes were small enough that the stiles and rails provided the necessary structure. The muntin configuration consists of seven pieces: one through-tenoned vertical muntin (or sash bar), and six stub-tenoned horizontal muntins (sash assembly drawing, facing page).

In traditional sash-making, the stub tenon extends only as deep as the sticking

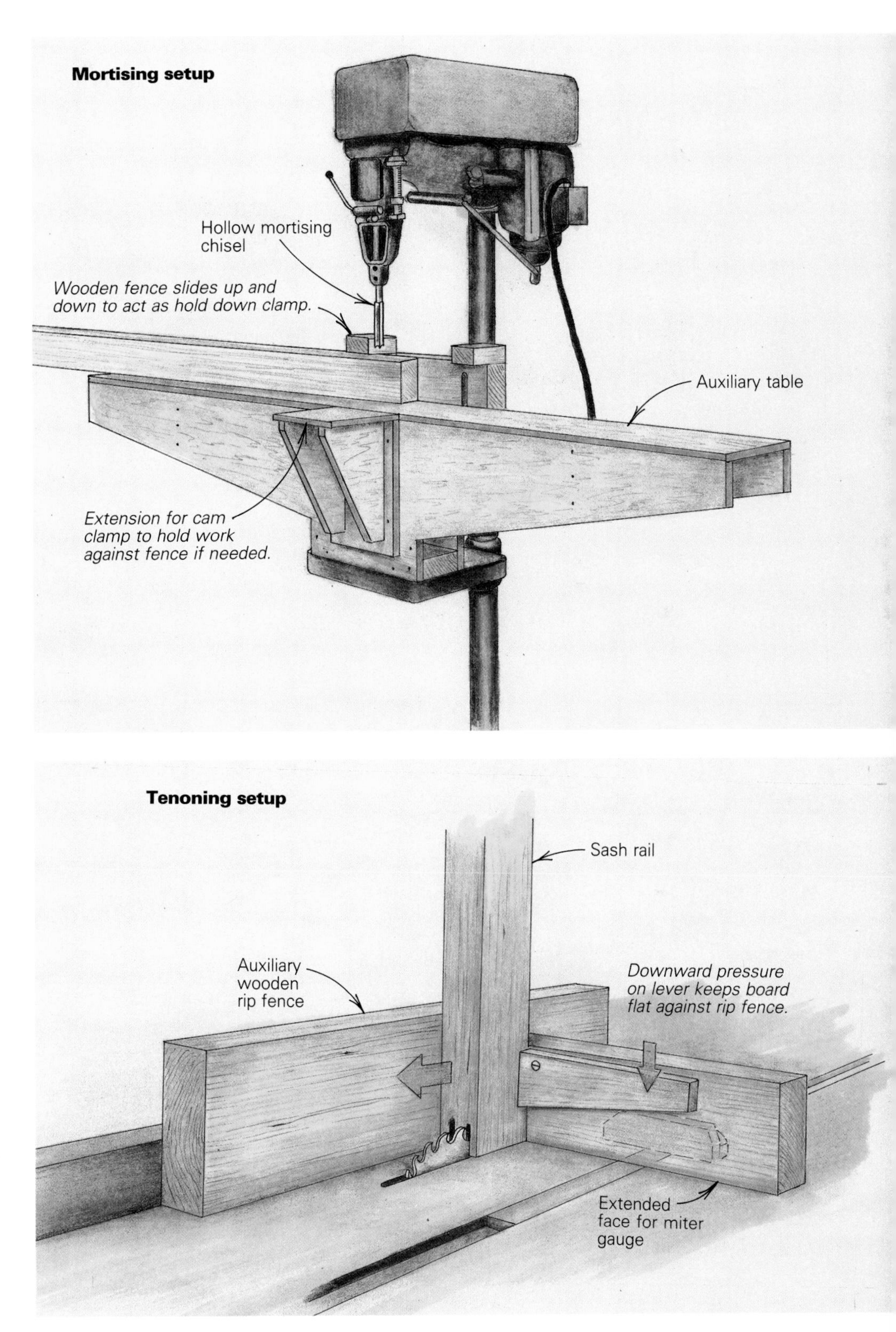

Drawings: Michael Mandarano

(for more on traditional sash-making, see *FHB* #18 pp. 72-77). Because I needed a bigger rabbet to make room for the insulated glass, I had less stock to house the stub tenons. Consequently, they had to reach somewhat further, extending into the face width of the adjoining member. At the vertical muntin, stub tenons would be reaching in from both sides. Because the face width of the vertical muntin is only ½ in., the stub tenon length had to be a little less than half that, or 3⁄16 in.

To finish the height rod, I labeled it with the word "height" on the top, then cut a point on this end to avoid confusing it with the squared-bottom end, which would be used to register the rod on the stock.

Layout of the width rod was essentially the same as for the height rod. I used through-tenons on the rails, so their length was determined by the overall sash width.

Fabrication—The first step in the actual making of the sashes was to size each piece to rough width and length. Then I jointed the pieces and ripped them to finish width. I saved the thin off-cut strips from the final ripping operation for shims. Varying in thickness from paper-thin to ⅛ in., they came in handy later for positioning the insulated-glass units in the sash.

After crosscutting each piece to finish length, I laid out the mortises. To speed this up, I transferred the mortise heights from the rod onto one of the sash pieces, lined that piece up with all the others like it and squared across the lot of them. Although you can cut mortises with a hand-held electric drill and clean them up with a chisel, I find it a lot faster and easier to use a ½-in. hollow mortising chisel chucked in a drill press. I didn't need to mark the width of the mortise because it was controlled by the fence setup on the mortiser. My vintage Delta home-shop drill press is a precise but willowy machine. To enable it to handle the long, heavy stiles, I built an auxiliary table out of plywood that extends about 4 ft. on both sides of the drill press (top drawing, previous page). A short wooden fence at the back of the table slides up and down on a pair of bolts, doubling as a hold-down clamp when the wing nuts are tightened. I added a small extension at the front of the table to accommodate a cam clamp for holding the work against the fence, although in most cases I found thumb pressure to be adequate for this.

I cut the tenons on a tablesaw. The combination of a high auxiliary rip fence and a miter gauge enables me to do fast, accurate tenoning without a $200 jig. Before cutting the tenon cheeks, I screwed a wood extension to the face of the miter gauge and attached a wooden lever to that (bottom drawing, previous page). The lever is made so that slight downward pressure on one end will hold the board I'm tenoning snug against the fence.

Coping and sticking—I'm lucky to have two shapers in my shop—a light-duty shaper with a ½-in. spindle and a two-speed heavy-duty shaper (both were made by Rockwell, now Delta International Machinery Corp., 246 Alpha Drive, Pittsburgh, Pa. 15238). In an old manual called *Getting the Most Out of Your Shaper* (originally published by Rockwell, reprinted by Linden Publishing Co., 3845 N. Blackstone, Fresno, Calif. 93726), I read up on the use of shapers for sash-making. Then, armed with matching cove-and-bead molding and cope cutters, I had at it.

In order for the tenoned parts of the sash to fit snugly against the mortised parts, they have to be notched to fit over the sticking (drawing below). This notching process is called coping, and it's the trickiest operation in sash-building, especially when the joinery is mortise and tenon. This is because a through-tenon on a rail or muntin won't clear the top of a standard shaper spindle sticking above the cope cutter. The solution is to use a stub spindle, which is shorter than a standard spindle and uses a countersunk machine screw instead of a nut to hold down the cutter. This allows the tenon to ride just over the cutter (drawing below).

Unfortunately, you need counterbored cutters to use with the stub spindle, and these don't seem to be available in carbide (probably because most sashes are dowelled these days, so cutting around a tenon isn't necessary). As it turned out, the Delta high-speed steel #09-137 counterbored cope cutter costs more than a similar cutter in carbide, but the carbide version isn't counterbored. A machinist counterbored my carbide cove-and-bead cope cutters (#43-915 and #43-916) so that I could mount them on the stub spindle to cope the shoulders of pieces with through tenons.

To form the stub tenons on the muntins, I used the same cutter on a regular spindle in conjunction with a spacer collar and a ½-in. straight cutter with shorter radius. This way I could cope the ends and form the tenons in a single pass.

Some sash builders cope the ends of all the muntins at once by cutting the coped shape across the end of a wide board and then ripping the individual muntins from it. I ripped each one to size first, and then coped them individually, using a sliding jig that Del-

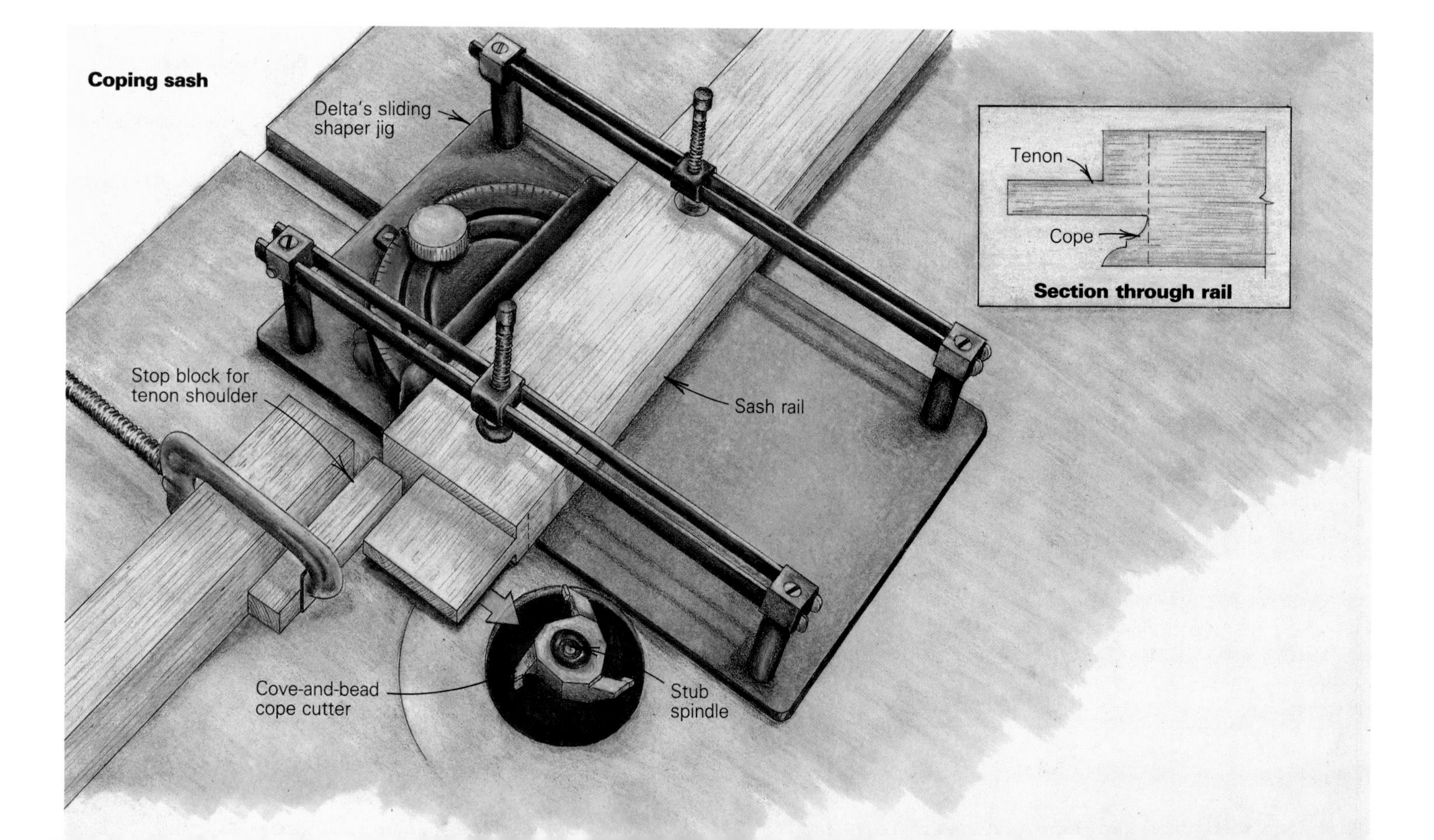

ta makes for their shapers (drawing, facing page). This jig combines a miter gauge and a small sliding table with integral hold down clamps. I coped the ends of the muntins before molding their edges so that I would have a flat surface bearing against the miter gauge. The subsequent edge-molding operation would remove the tearout left by the cope cutter. Sizing muntins individually lets me use up a lot of otherwise useless short and narrow scrap, as well as making it easier to account for all the necessary pieces before moving on to the next step.

To ensure perfect mating of cope and sticking, I set up both operations simultaneously on separate shapers. If I only had one shaper, my first step would have been to make an accurately fitting pair of prototypes by trial and error (one stuck, the other coped). These would have been used to test the various setups. The stiles and rails were beefy enough that I felt I could run them safely on the shaper without hold-downs. For the muntins, I used Delta's nifty spring hold-downs, which held the small molding securely to the fence and table.

A wide straight cutter could have been used below the sticking cutter to do the rabbeting at the same time, but it would have meant making a heavier cut than my equipment is comfortable with. Instead, I used a narrow straight cutter along with the sticking cutter. This ploughed out part of the rabbet. I sliced off the remaining material on the tablesaw.

The slotted track and crank housing of the casement operator are simply screwed in place. The "crank molding" is the narrow stock notched to fit over the crank housing. It provides a surface for the screen to sit on. The solid-brass hinges have a small set screw in the barrel that prevents the hinge pins from being removed from the outside.

Assembly and installation—After all the shaper and saw cuts were complete, the sashes were ready to assemble. The nice thing about through-tenoning is that it allows me to start all the tenons in their respective mortises. Then all I have to do is brush a light coating of glue—resorcinol in this case—into the mortise from the opposite side and tighten the clamps. That saves a lot of frantic moments. I used clamping blocks that have a channel ploughed in one side. Centering this channel over the mortise before tightening the clamps allows the tenon to protrude slightly beyond the edge of the stile if necessary.

After laying the insulated-glass units into the sashes and shimming them into position, I was faced with the task of fitting all the moldings that would hold the glass in place. One of the rooms I worked in, for example, contained six sidelights, five casements and one 8-ft. pair of French doors. That translated into 112 panes of glass, which meant 448 moldings requiring 896 miters. There had to be a better way.

My shortcut was to cope wide stock on the shaper first, then mold the sticking on the edges and rip off individual precoped moldings. This trick saved many tedious hours hunched over a screaming miter box. Another time-saver I discovered was a pneumatic brad tacker. It came in mighty handy for the 2,688 fasteners.

Hanging the sash was no different from hanging doors. I used a router and a plywood template to cut the hinge mortises in the sash. On the new jambs that I made, I had routed mortises for the hinges before assembling and installing the frames.

Because the hinge barrels for each sash would be located on the outside, I wanted solid-brass butt hinges to resist corrosion. I also needed the security of a nonremovable pin. The ones I used were made by the Baldwin Hardware Corp. (841 E. Wyomissing Boulevard, Box 15048, Reading, Penna. 19612) and cost $50 a pair. The end caps on the barrel of the hinge unscrew, allowing removal of the hinge pin for easier installation of the sash. A tiny set screw on the indoor side of the hinge barrel secures the pin afterwards. The casement locks were simply screwed to the sash and their strike plates drilled and mortised like those for the lockset on a door.

I worked out the location for the casement operator by trial and error. If the crank was too close to the jamb, there wasn't enough room for a hand to turn it. And if the crank was too far from the jamb, the window wouldn't open very far. Once I figured out the best location, installation was straightforward—a slotted track was screwed to the inside face of the sash and the body of the casement operator was screwed to the window stool. I made a notched wooden piece, which I call a "crank molding," to fit over the casement operator (photo left). In addition to being notched for the body of the casement operator, it is rabbeted in the back to receive the retracted arm of the casement operator when the sash is closed. There's also a shallow rabbet in the top for the screen to sit in. I made all the cuts on the tablesaw, starting with the cross-grain notch for the operator.

Weatherstripping the windows was a prime concern. I chose a stainless-steel leaf-type material called Numetal, made by Macklanburg-Duncan (Box 25188, Oklahoma City, Okla. 73125). Each box of Numetal includes nails and about 30 ft. of weatherstripping. I nailed it to the side and head jambs along the sash rabbet. It is fairly easy to apply, invisible when the sash is closed and is the most durable of any system I've used.

On the bottom of the sash, I used a rubber sweep on the outside to seal the sash against the sill. Then I mounted a piece of compressible gasket-type weatherstripping on the stool for the sash to press against. I could have run it along the sides and head as well, but the material would have interfered with the fasteners on the strike side and might have gotten pinched by the closing sash on the hinge side. It wouldn't have looked very good, either.

For the screens, I built simple frames out of ¾-in. stock. I cut a rabbet on both sides—a shallow one on the inside for decoration and a deeper one on the outside for the screen and screen bead molding. I coped the rails into the stiles, which essentially creates a shallow bridle joint, glued them together and ran a long screw through the ends. I cut a narrow screen-retaining groove in the bottom of the rabbet on the screen side and used a screening tool (which looks like a dull pizza cutter) to press in the aluminum screen. I used a utility knife to trim the screen against the corner of the rabbet and then nailed on the screen bead molding. To hold the screens in place, I installed bullet catches between the stiles and jamb—one on each side, about halfway up the stiles. A delicate brass knob, installed on the face of the frame, makes the screens easy to pop out in the fall for a clearer view. □

Scott McBride is a carpenter in Irvington, New York, and a contributing editor of Fine Homebuilding.

Building Fixed-Glass Windows

Working on the job site with the tools at hand, you can easily beat the cost of special orders

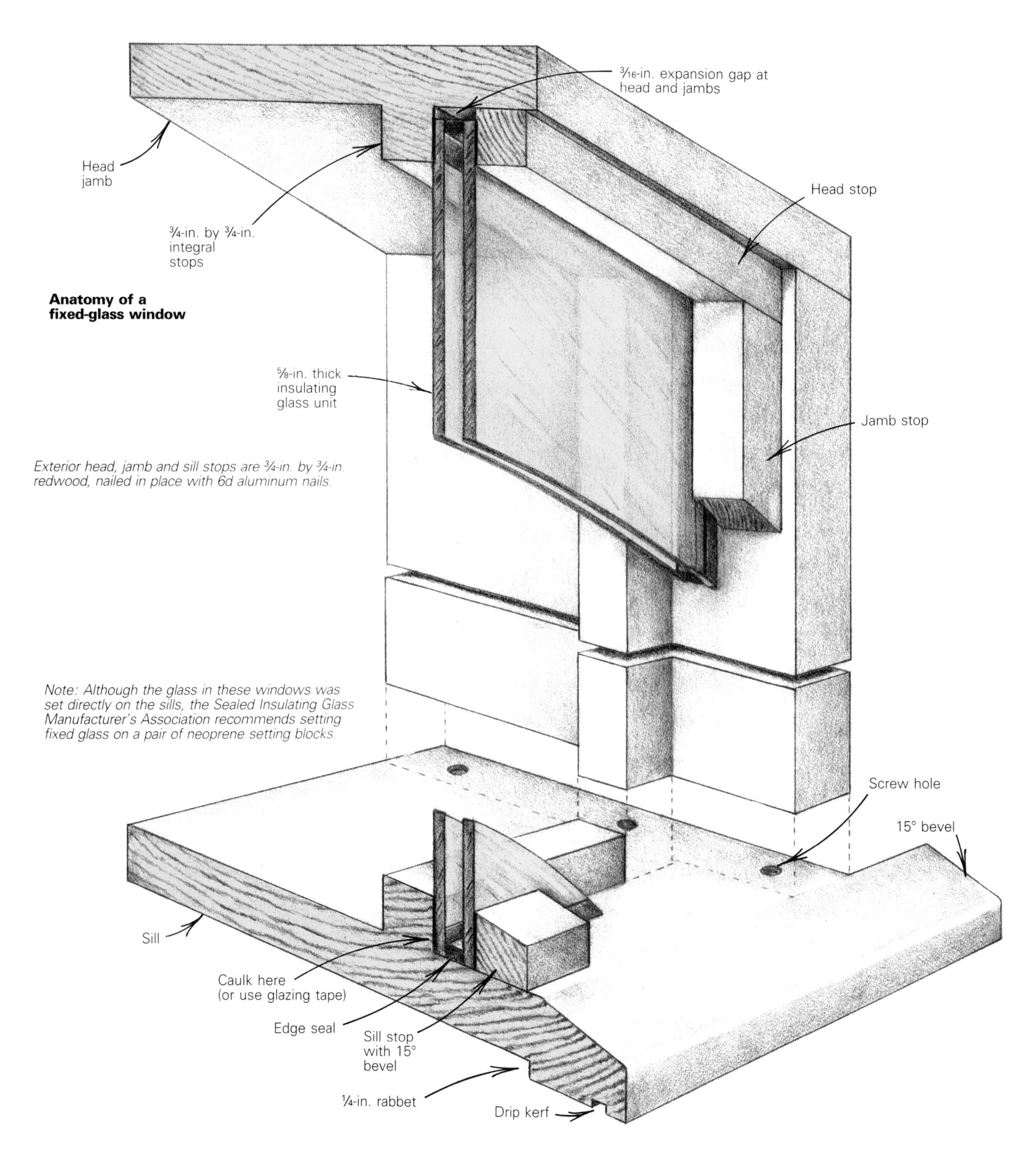

From *Fine Homebuilding* magazine (February 1989) 51:76-78

While building a house a few years ago, we experienced the usual number of surpises at material costs and delivery times. One of the worst surprises concerned fixed-glass windows. The home is a passive-solar design, so it includes many such windows (photos below). We reasoned that because these windows were substantially simpler in construction than the wooden casement windows used elsewhere, they would cost less.

A typical wooden casement is a marvel of precision construction, maintaining airtight weatherseals over many linear feet of sash and through years of winter storms and summer heat. In the San Francisco Bay Area, casement windows typically cost $20 per sq. ft. of glazed area, including screens. Although window wholesalers told us not to rely on this approximation, we found it to be generally accurate.

We were shocked to discover that a fixed-glass equivalent would run $35 to $40 per sq. ft. of glazed area—an assembly without moving parts or weatherstripping. Adding insult to injury, the first company we contacted quoted a 14-week delivery time. Other companies promised slightly better delivery times, but couldn't break the $35 per-sq.-ft. price barrier.

We did not consider ourselves window craftsmen. And it remains true that casement windows require so much special tooling that we could not reasonably compete with a production window shop (see the article on pp. 39-43 for more on making casement windows). But Tom Hise, the project architect, convinced us that we could build quality fixed-glass windows in a reasonable amount of time and do so at a price (including our labor, of course) far below the alternatives.

Simple frames—The house is a single-story contemporary with opposing shed roofs and a clerestory. Because the fixed-glass windows follow the roof lines they are trapezoidal. Altogether there are 12 of these windows scattered around the house. We started by laying out all the windows full-scale on the subfloor.

By far, the most important detail in any window is the sill. The sill is the last stopping point for water before it either drips harmlessly off the house, or is pulled destructively into the walls by capillary action. The 15° bevel on the front of the sill directs water away from the building, while the drip kerf underneath the sill guarantees that capillary action won't pull the water into the house (drawing facing page). Milling the jambs and sills from solid stock and incorporating integral stops (rather than using applied stops) similarly prevents water from migrating past the glazing.

The window sills are 1¾ in. thick altogether, with a ¾-in. thick integral stop, a ¾-in. thick center section and a ¼-in. rabbet in the bottom that fits over the rough opening and further discourages water from entering the house. The head and jamb pieces, which are identical to each other in cross section, are essentially sill pieces without the bevel and kerf details, but they're only 1½ in. thick altogether because they don't need the ¼-in. rabbet. The jambs and sills are made of #1 clear white pine and were milled on a tablesaw with a dado blade.

All the surfaces that would show after installation were first sanded, and then the pieces were cut to the proper length and angle. The integral stops on the end of each piece had to be cut back (notched) in order to butt the side jambs into the headers and sills (drawing facing page). We cut the sills so that the beveled portion extended past the side jambs on both sides by the width of the exterior trim. The frames were glued and screwed together, then each was laid on top of its respective chalk-line template on the subfloor to check the dimensions.

Setting the frames—After the glue had cured, we painted the window frames, the exterior stop and the exterior trim with two coats of primer. After cutting the exterior trim, we attached it to the jambs and heads with a pneumatic finish nailer, driving 8d aluminum finish nails. We stay away from electroplated galvanized nails because they seem to lose their plating and eventually bleed. If we were hand-nailing, though, hot-dipped galvanized 8ds would have been fine. The nail holes were puttied and sanded smooth.

The finished (but unglazed) frames were then set in the rough openings, plumbed and nailed into place. We use standard flashing details (see *FHB* #9, pp. 46-50), but for good

The window configuration shown above is repeated twice on the rear elevation. So when the window companies quoted a price per sq. ft. for fixed-glass windows that was nearly twice that of the casements, the authors decided to build their own fixed-glass windows.

Building fixed-glass windows is simpler than you might think. After making and installing the window frames, a bead of caulk is run around the interior stops (left) and the double-pane insulating glass is set in the opening (right) and held in place with wooden stops.

Drawing: Michael Mandarano

measure, we always run a bead of caulk between the top of the trim and the building paper. After installing the siding, we caulked again between the siding and trim.

Fitting the glass—To make sure the insulating glass would fit into our frames, we cut cardboard templates ⅜ in. smaller than the width of the window opening and $^{3}/_{16}$ in. smaller than the height and gave them to a glass company. When installing the glass, you can either seal it with glazing tape or with caulking. But if you use caulking, make sure it is compatible with the seal used by the insulating-glass manufacturer. In this case, we used caulk and applied it against the vertical face of each integral stop (photo previous page, bottom left). We followed with a second application on the outside between the glass and the exterior stops. We used redwood for the exterior stops and nailed them up with 6d aluminum nails. For extra protection, we cut 15° bevels in the sill stops.

We set the glass directly on the sill and have had no problems with it in the two years that the windows have been in place. We've since learned, however, that the Sealed Insulating Glass Manufacturer's Association (SIGMA, 111 E. Wacker Dr., Suite 600, Chicago, Il. 60601) recommends setting fixed glass on a pair of small neoprene blocks (called setting blocks), which help distribute the weight of the glass and prevent water from being trapped behind the glass. SIGMA also recommends drilling a pair of weep holes in the exterior sill stops.

Corner sidelight—At the front entrance of the house we built a large sidelight with two panes of glass meeting at right angles (photo below). In this case only one header—supported by an exterior wall on one end and by an interior partition on the other—was needed to carry the roof loads. It is conceivable that two headers could be required in circumstances where two load-bearing walls intersect at the window. Several manufacturers of metal connectors make a framing clip for headers that intersect other headers. Because of the unusual glazing detail, it was critical that the rough opening be plumb on either side of the window's corner.

The corner unit was built with the same jamb, sill and head sections as the other windows. First the sill and jamb stock were fabricated as described in the previous sections. Then the sills and heads were mitered and cut to length in matching pairs. Accuracy in cut length was important to guarantee a square opening for the glass. The jambs were cut to matching lengths and the whole unit was assembled near the rough opening.

Before the glue had a chance to set, we placed the unit in the rough opening, aligned the corner of the sill with the corner of the framing, shimmed it level and tacked it in place. We used a plumb bob to align the mitered corner of the head jamb to the identical point on the sill below. Next, the sill was tacked near each jamb and the head adjusted in or out until plumb. Then we stepped back and double-checked that everything was plumb and level.

It was not critical that the corner be exactly 90°. The critical requirement was that the jambs were plumb so that the two panes of glass would meet neatly at the corner. If small adjustments were needed, this was the time to make them. Once everything was plumb and level, we set the nails, puttied the holes and sanded them smooth.

Because of the proximity of the window to the door and to the floor, we had to use safety glass. And to achieve a clean line at the intersection of the glass, we used single panes (¼ in. thick) rather than double-pane insulating glass, which would have made an awkward corner. We installed the glass exactly as before except that we applied a bead of clear silicone between the mating glass surfaces at the corner. After the silicone set, we trimmed off the excess inside and out with a razor blade. The final step was to miter the exterior stops and nail them into place.

Number crunching—When the windows were finished, I calculated how much they had cost us. We paid $4.34 per sq. ft. for the insulating glass, $3.29 per sq. ft. for the tempered glass and billed our time at $40/hr. The cost for all 12 windows and the corner sidelight averaged just under $15 per sq. ft. □

Because it is close to the door and the floor, this corner sidelight had to be made with safety glass. The corner glazing detail is simply two sheets of glass butted together with a bead of clear silicone between them. The drywalled header that carries the roof loads over the window is visible at the top of the photo.

Jay Chesavage is a contractor and engineer in Palo Alto, California. Steven Tyler is a cabinetmaker and carpenter in Concord, California. Photos by the authors.

A Site-Built Ridge Skylight

How an architect built an acrylic skylight for his remodeled kitchen

by David K. Gately

Large residential skylights have always posed problems for architects, contractors and owner/builders. A big custom-manufactured skylight, with tempered or wire glass, extruded-aluminum frame and all the necessary gaskets can be costly. It's also hard to install without a crane and is visually cumbersome—if not downright ugly.

But I think the benefits of a generous skylight far outweigh the disadvantages. It brings the sky, rain, clouds and trees close enough to appreciate. Skylights also serve to reduce the effects of window glare by balancing the natural light in all parts of a room.

I've had some experience fabricating acrylic covers for architectural models, and the thought occurred to me that I could apply the same material and technology to the problem of assembling a site-built skylight. When it came time to design and build my own kitchen addition, I decided to try out a skylight design I've had in the back of my mind for some time. As shown in the drawing on the next page, the design is based on modular panels that interlock, with the flashing edge at the end of one panel nesting in a U-shaped gutter in the adjacent panel.

My case was special in that our house in Mill Valley, California, is well above the road—75 vertical ft. and 135 steps. I needed a skylight material that was light and easy to handle and did not require special technology to fabricate or install. No cranes would come a-calling.

The space that is skylit is a country kitchen 28 ft. long by 15 ft. wide (photo, p. 51). The ceiling is 1x6 cedar decking supported by 2x6 cedar trusses with a roof pitch of 6 in 12. The ridge skylight is 22 ft. long, spanning six 3-ft. 8-in. wide bays (photo below). I wanted to create the most unobtrusive skylight possible and one that also would solve all the problems of a field-built skylight—namely differential movement of dissimilar materials such as wood, metal and plastic.

Thermal expansion—Movement caused by temperature change is the first consideration in designing an acrylic skylight. The coefficient of expansion of acrylic is .000034 in. per degree of temperature change. This means that in a climate with a yearly temperature range of 20° F to 90° F, the skylight must be designed to accommodate a minimum change in the dimension of each element of 70° F x .000034 in. of expansion x the dimension of the skylight panels. Certain parts of a skylight will heat up well beyond the air temperature. It is wise, therefore, to design for a temperature differential of 100° F in a moderate climate such as ours.

My panels are 44 in. long and 37 in. wide. This means that I must accommodate a length-

The six interlocking segments that compose Gateley's skylight extend 22 ft. along the ridge of the roof. Sunshade cords dangle over acrylic support bars just below the ridge of the skylight. In the foreground you can see the acrylic cap piece and gusset that reinforce the panels.

wise dimensional change of .15 in. (44 in. x 100° F x .000034 in.) and .13 in. widthwise in each panel. To be on the safe side, I designed this skylight for about ¼-in. expansion of each panel in each direction.

Hold-downs—Each panel moves independently with relation to its neighboring panel. The tricky part is attaching the panels to the curb so that they are secured to the roof, yet allowed to move.

My strategy was to use some L-bolt hold-downs that I fabricated in the shop out of ⅜-in. and ¼-in. brass threaded rod (section drawing facing page). The ⅜-in. leg of the bolt extends about 2½ in. into the skylight curb, and the ¼-in. leg fits through the overlapping edges of two skylight panels at the gutters. My skylight has been installed now for a year and a half and has been subjected to strong winds, rain and the full range of temperature change. The hold-downs work well, but I think they are overkill except for skylights that are exposed to high winds. Also, they are tedious to install.

The next time I build one of these skylights, I'll use the L-bolts only at the corners. To link the field panels, I'll use hold-downs that are made of three built-up layers of ¼-in. acrylic. I used some at the bottom of the gutters where they pass over the curb (drawing facing page) and they work well. They are attached to the curb with a bolt made out of ¼-in. brass threaded rod. Above the acrylic hold-down, a ¼-in. brass bolt through the overlapping gutter and flashing edges loosely ties together the adjacent panels. The main point here is that each skylight segment, which is composed of two panels rigidly joined at the peak, must be able to move freely in all directions about ¼ in.

The house is in a relatively protected location and the skylight is not subject to strong winds. A hilltop or ridge location would require additional thought for counteracting 100 mph winds and the resulting uplift. The two hold-down methods described here would suffice for all but the most exposed locations.

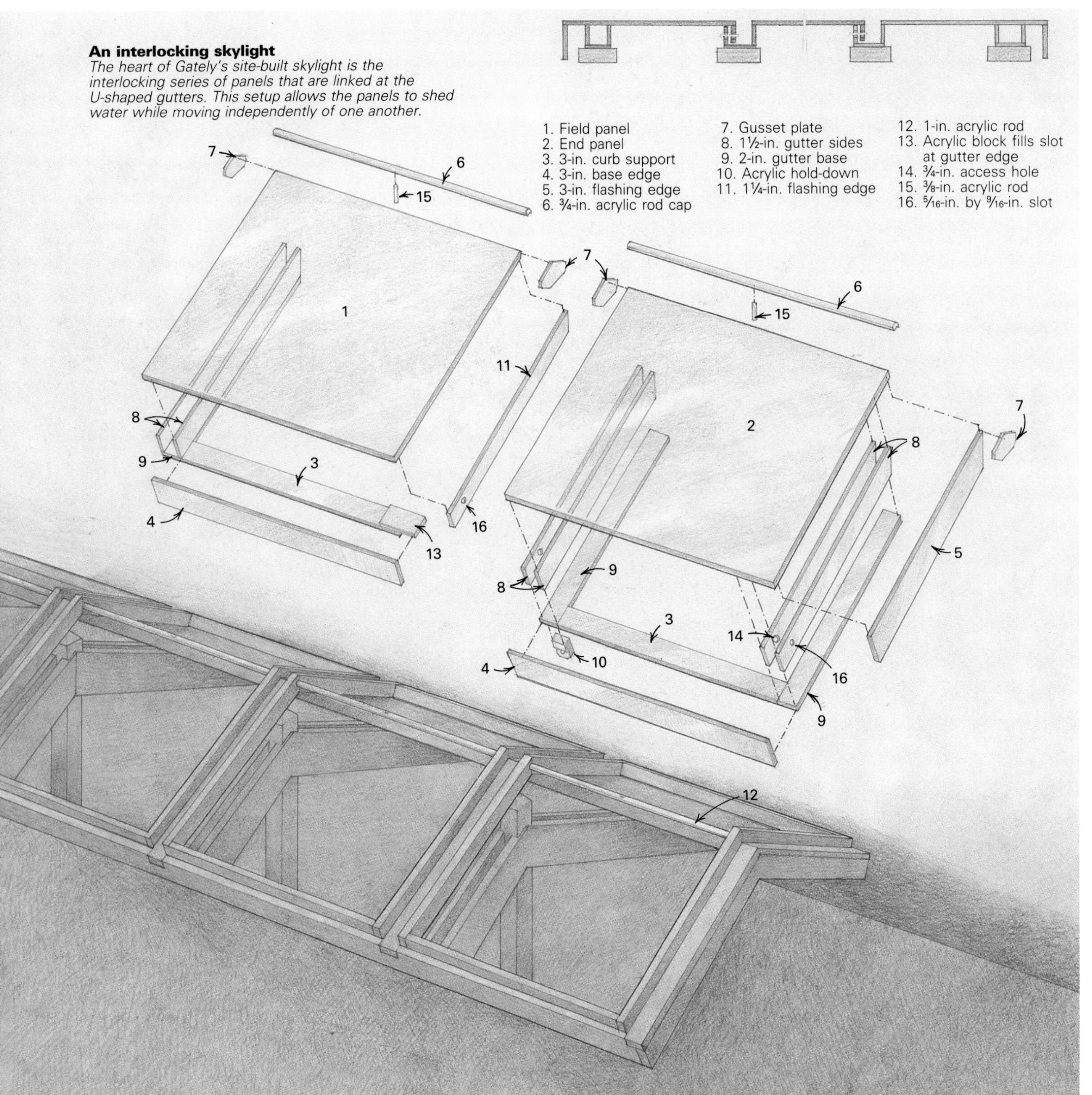

An interlocking skylight
The heart of Gately's site-built skylight is the interlocking series of panels that are linked at the U-shaped gutters. This setup allows the panels to shed water while moving independently of one another.

From *Fine Homebuilding* magazine (August 1989) 55:45-49

Shop drawings—Whether you farm out the acrylic cutting or do it yourself, a descriptive drawing of each component, listing the precise number of pieces required, is essential. A shop drawing need not be to scale and can be done freehand, but it must be precisely dimensioned and clearly marked as to hole locations and bevels. It helps to draw each component on a separate sheet of paper. List the number of pieces to make, assign each one a letter designation, and show a small master-plan view of the whole skylight on each sheet, showing with a darkened line where each piece is located.

Components—The plastic I used was ¼-in. thick cast acrylic with a 10% grey tint. The brand is "Acrylite," which is similar to the brand "Plexiglas." I chose, in the interest of time and because I did not have an adequate table-saw setup, to have the components cut by a local plastics supplier (City Plastics in San Francisco). Only the side pieces terminating at the peak, where they are joined with an opposite panel, need to be precisely cut. Otherwise, the materials are pretty forgiving, and any small discrepancies can be dealt with during installation.

If I had had a good table saw and adequate shop space, I would not have hesitated to cut the components myself, thus saving about $600 for about a day's work. The major tools required are a table saw with an 8-ft. outfeed table, a plastic-cutting carbide planer blade with as many teeth as possible and, for beveling the panels that meet at the peak, a router fitted with a custom bit that matches the plumb cut for the roof pitch (this will cost about $30). Straight, parallel cuts are essential; absolutely smooth cuts are not, although that would be ideal. The protective paper should be left on during all cutting and beveling. Before you take router to acrylic, be sure to make lots of practice cuts.

Drilling the attachment holes—The holes through which the brass bolts pass should be slotted, as it is here that all the expansion of the material is accommodated. For a slot 5⁄16 in. wide by 9⁄16 in. long, drill two ⅛-in. starter holes with their centers ¼ in. apart. Then drill each hole partway through with a 5⁄16-in. bit, alternating each hole as you go to keep the two holes centered. Keep the protective paper on the plastic as you drill, and clamp the workpiece to a waste piece of thick plastic. Practice drilling a few holes first to get the feel of drilling this material. Acrylic has a tendency to shatter as the bit penetrates the opposite side unless you reduce drill pressure as you finish. Incidently, there are special drill bits designed for working with acrylic. Each has a pointed tip that reduces the likelihood of shattering the material. Plastics suppliers sell them.

Panel assembly—Drilling complete, strip the protective paper from one set of panel components. The only surface left protected is the upper side of the skylight panel. Keep the paper on this surface until the panels are on the roof as this will allow you to slide the panel around the fabrication table and up the ladder.

I assembled the 1¼-in. flashing edges using large spring clamps, and used bar clamps for the 3-in. flashing edges. Once a piece was clamped in position, I applied a solvent called Weldon #4 (IPS, P. O. Box 471, Gardena, Calif., 90247) with a hypoapplicator (a polyethylene bottle with a syringe tip) along the length of the joint. It's easy to see the spread of the solvent as it moves through the joint via capillary action. Once you've flooded the joint, release each spring clamp briefly to allow the liquid to penetrate the joint fully at that point. Less clamping pressure will result in some excess dissolved plastic protruding from the joint but will produce a stronger joint. Too much pressure will result in a cleaner but slightly weaker joint. This is not critical, but a clear, bubble-free joint is desirable. The process takes about ten minutes per part: three minutes to set up and clamp, one minute to apply Weldon #4 and about four to six minutes for setting and minor adjustments. One or both hands act as additional clamps at this time as you are constantly examining the joint for a clean, transparent set. If you miss a spot, you can always apply more Weldon #4 to achieve a fully welded joint.

Use care with the Weldon #4 liquid, but don't worry if it gets on plastic where you don't want it. Just don't touch spills with anything because that can cause scars. Spills dry rapidly and leave a thin white residue that cleans up with Meguire's Mirror Glaze (Meguire's Inc., 1 Newport Pl., Newport Beach, Calif. 92660).

Assemble the gutters and set them aside. Alignment is most critical where they will be joined at the peak. If they don't quite line up, any overage can be sanded off; underage can be filled during installation with a thick glue called Weldon #16 cement (more on this product later).

Now assemble each field panel, beginning with the 1¼-in. wide flashing edge. Next glue on the 3-in. wide base edge, followed by the gutter assembly. The last piece to be glued in place is the 3-in. wide curb support strip

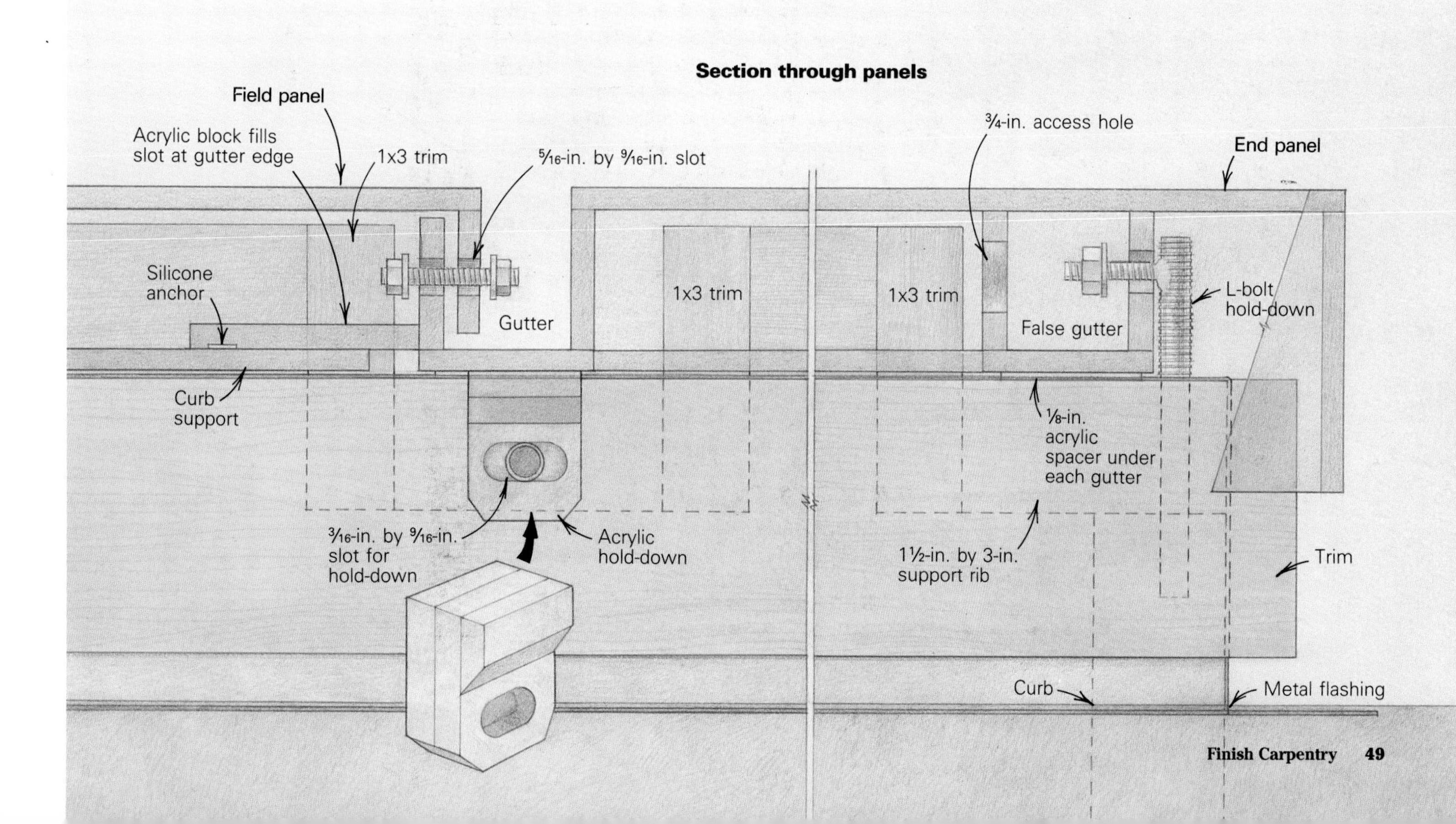

(drawings, pp. 48 and 49). It abuts the gutter at one end of the panel, but stops 1 in. shy of the flashing edge at the opposite end of the panel. The resulting notch allows the panel to be fitted over the interlocking edge of the gutter. To position the curb support piece, scribe a fine line on the base edge using a razor knife guided by a straightedge.

The field panels each have a gutter along one side and a flashing edge on the opposite side. The end panels, however, are a little different. They have a 3-in. wide flashing edge that runs from the base edge to the peak. Also, they have a "false gutter" that rests atop the curb assembly at either end of the skylight well (drawing previous page). The false gutter acts as a box beam to strengthen the panel, and as a spacer to keep everything in the right plane. The right end panel has a gutter assembly on its left side. The left end panel has a flashing edge on its right side.

When all the panels are assembled, hand-sand any parts of the gutter assemblies that project past the peak. I sanded all the edges that meet at the peak, including the bevel on the panels, to make these pieces even. A sanding block with #80-grit paper contact-cemented to a thick piece of scrap acrylic works well.

Installation—With the paper still on the upper face of the plastic, move the panels onto the roof. I just pushed them up the ladder, paper side down. Although they are not fragile, one must take care not to torque them, which may loosen a joint. If this does happen, just apply some more Weldon #4 to the separated joint, and hold it steady until it sets up again. Each panel weighs perhaps 20 lb. and is slightly cumbersome, but it is still just a one-person job to move one.

At this point place the ⅛-in. thick acrylic spacer/bearing strip on the tops of the support ribs. I used small finish nails along the sides of the ribs to retain them, leaving just enough of the nail head exposed to catch the edge of the acrylic strips. Now is also the time to affix weatherstripping to the metal flashing atop the curb (drawing below). I used the self-sticking neoprene foam type, ⅜ in. wide by ¼ in. thick.

Attachment—The panels are anchored with the L-bolt hold-downs at each corner. At the end panels, the ¼-in. dia. leg of the hold-down extends into the false gutter, where it's reached from inside the skylight well through a ¾-in. hole drilled through the inboard edge of the false gutter. The brass nut and washer are installed through this access hole. I put a little contact cement on the washer, which temporarily stuck it to the socket during installation.

I began the installation with the right-side end panels. With the two panels sitting in the correct position, their peak edges aligned, I clamped them together with several elastic cords linked together, stretched over the peak and hooked to the base edge of each panel. Panels aligned, I marked the curb for placement of the L-bolts at both leading and trailing edges of the panels, leaving enough space between panel and bolts for longitudinal expansion of about ¼ in. Transverse expansion is accommodated by the slotted holes. Then I removed the panels, drilled my holes and installed the L-bolts.

The horizontal leg of the L-bolt must be short enough to allow the panel to be threaded onto it, but long enough to accept a washer and nut (about ⅝ in.). The panels are still unjoined at this point so there is a lot of play to accomplish this maneuver. The L-bolts can also be turned a little to help positioning. To align some of the bolts, I reached through the access holes with a hooked wire.

Once I had the first two end panels in place, the rest of the panels were quite easy to position. At this stage it is a simple matter to adjust the panels for alignment by finger-tightening the nuts from the outside while kneeling on the roof. Don't be alarmed if the ridge edges of the panels sag when first placed. They will become straight, rigid and structural when joined with Weldon #4 glue.

Once I had all the panels in place, I removed the protective paper. This is a nice moment of revelation because the light is finally allowed to stream into the room. But the skylight is not watertight yet, and in this case, with night coming on, I ran duct tape down the ridge in case of rain. Rain it did, and the tape kept it out.

Joining the ridge—Working from the outside, I used two trios of elastic cords to hold opposing panels in place at their ridge ends during glue-up. To prevent scratches, I wrapped the hooks on each cord with duct tape.

I used an awl inserted vertically between the panels to adjust the alignment of their edges. If one of the panels is a little low, lifting on the awl brings it back in line. With the ridge and panel ends precisely aligned and under tension from the elastic cords, I applied Weldon

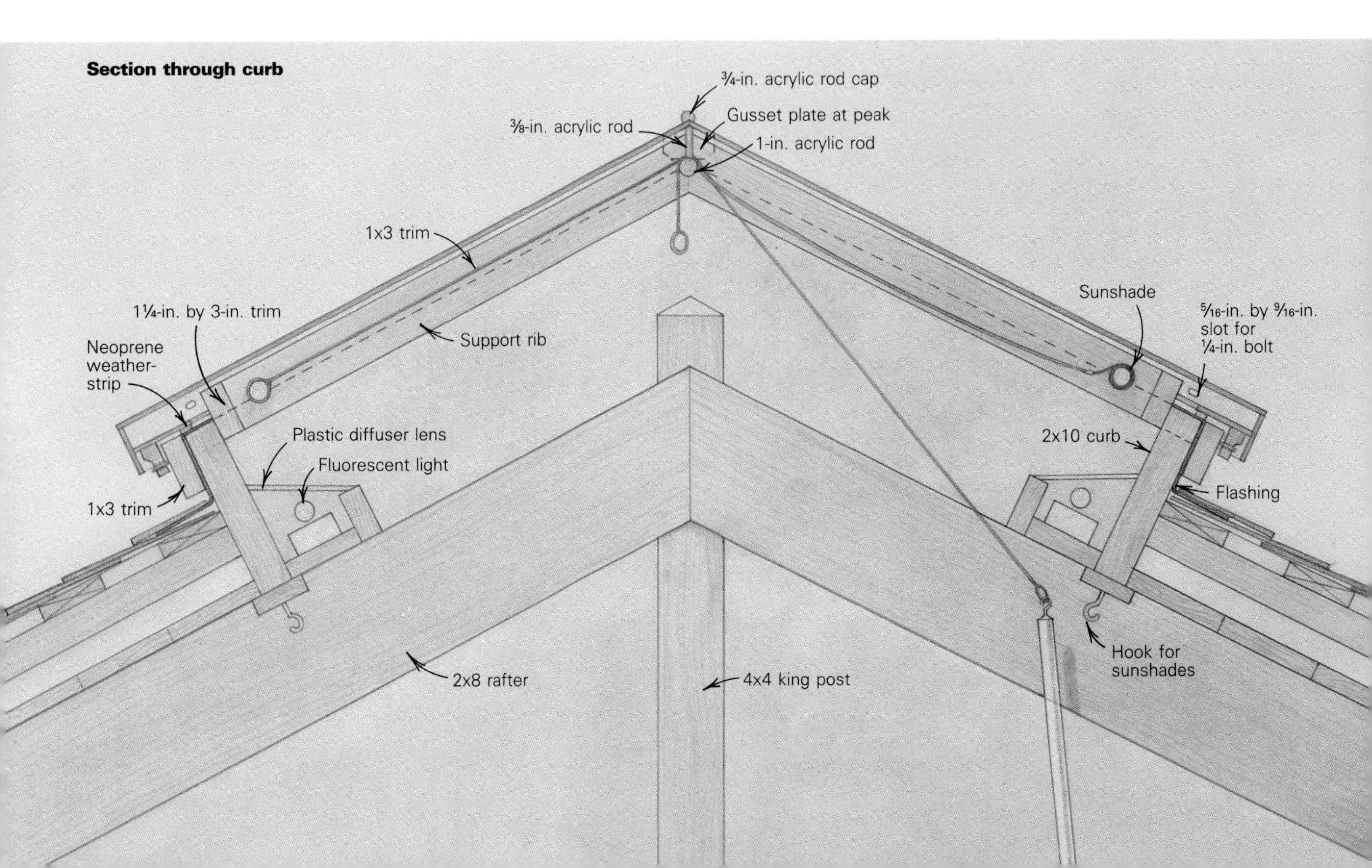

#4 liquid to the ridge joint with the hypoapplicator. I started at the center of each panel and worked outward, flooding the joint with the solvent as far as I could reach. Then I pressed the ridge joint together with my fingers and held it briefly until it took a set. I didn't worry about joining the entire length all at once, and I found that a few inches of the joint at each end wouldn't entirely close. I filled these gaps with Weldon #16 clear cement. This is a thickened, syrupy version of Weldon #4 used for high-strength joints that need some filler. This cement flows fast, so crimp the nozzle to 1⁄16 in. or less to reduce flow.

After each succeeding pair of panels was joined with Weldon #4 and the ends were filled with Weldon #16 cement, I applied a small bead of Weldon #16 to the outside and inside of the skylight peak along its entire length. This is really the only part of the gluing process requiring any degree of skill. A little practice reveals the behavior of the cement under vertical and horizontal conditions; it hardens at a certain rate and flows from the tube evenly if applied at an even speed. Uneven flow can be corrected while the cement is still liquid by backing up slightly and stretching out the bead while slightly increasing the flow. An even, 1⁄8-in. wide bead is the goal. Weldon #16 will occasionally whiten as it cures. This happens unevenly, so use it sparingly where it shows from below.

To finish the installation I capped the notches that remained in the curb supports (next to the gutter assemblies) with a square of 1⁄4-in. acrylic, secured with a dab of silicone caulk (drawing, p. 49). Next time I trimmed out the support ribs with 1x cedar. To clean up the scuffs and dribbles of installation, I polished the main panel surfaces both inside and out with Meguire's #17 cleaner, followed by an application of Meguire's #10 mirror glaze.

The test of summer—When I initially discussed the structural feasibility of my ridge joint with the plastic supplier, we both thought that a plain butt joint would be adequate. In the event that it was not, a ridge cap could be installed to solve any structural problems that might arise. Sure enough, during the hot summer weather, each opposing pair of panels expanded and contracted so actively that the unreinforced ridge joint gradually failed, starting at the ends and working inexorably toward the middle.

To counter the forces involved, I added gussets to the vertical gutter sides at the ridge (drawings, p. 48 and facing page). To further strengthen the ridge and create an absolutely watertight joint, I added a cap made from a 3⁄4-in. acrylic rod V-grooved lengthwise so it conforms to the shape of the peak. These solutions have worked well, and should be considered an integral part of the design. They also eliminate the need for the meticulous application of Weldon #16 cement.

An inevitable by-product of a clear skylight is the summer sun's effect on things such as candles, butter and fabrics, namely wilting, melting and fading. I had planned from the start to install some form of shading device, and living with the sun hastened the need.

I settled on a design that complements the lines of the acrylic cap piece. A 1-in. dia. acrylic rod is now located in each skylight bay, 1½ in. below the peak (drawings, p. 48 and facing page). The rod is let into and supported by the wood trim. It flexes slightly under the pressure of drawing the shades, so I installed a 3⁄8-in. vertical rod at the center of the span. It is let into 1⁄2-in. deep holes drilled into the ridge cap and the 1-in. rod, and it's affixed with Weldon #4. The shades are mounted to the sides of the trim pieces (drawing facing page), and their pulls are draped over the 1-in. acrylic rod. To close the shades, I use a 7-ft. pole with a hook at the end and slip the shade pulls over hooks screwed into the ceiling trim.

The cost—I priced a custom metal and glass ridge skylight at $3,500, single-glazed, not including installation which would have been costly—probably another $1,000. My cost for acrylic materials was about $1,000 for 132 sq. ft. of skylight, or $7.50 per square foot. Fabrication took me about 20 hours and installation about 24 hours, which included plenty of head-scratching as I puzzled out the assembly process.

The whole assembly of skylight and shades is now fully operational and I anticipate years of trouble-free service. Since the acrylic panels are so active, especially during the summer, I plan to inspect the edge joints on the panels when I make my annual autumn trip to the roof to clean the gutters. If I find any joints that show signs of opening, I know I can reseal them with a squirt of Weldon #4 from the hypoapplicator. □

David K. Gately is a partner in the architectural firm of Callister Gately Heckmann & Bischoff in Tiburon, California.

Rows of red cedar trusses are illuminated by the ridge skylight in the author's kitchen. To keep the room from overheating on hot days, Gately installed retractable shades near the skylight curbs. They can be opened and closed manually with a pole that has a hook mounted on the end.

Drawings: Gary Williamson

Installing Arch-Top Windows

How one builder supports the loads without a conventional header, then uses a trammel jig to cut siding and casing

by Douglas Goodale

Arch-top windows have caught the fancy of home buyers, so carpenters have to contend with the difficulties of framing and finishing these semicircles within rectilinear walls.

Round-top, circle-top or arch-top, whatever you call them, such windows have become a popular architectural feature. They add light and view and a touch of grandeur to a house. But these windows require carpenters to come up with some resourceful solutions when it comes to installation and trimming. Here, I'll describe techniques that work for me, including some I used on a recent project in rural Hunterdon County, New Jersey (photo above).

Solving the structural problems—A typical rectangular window has a structural header at the top of the rough opening, and the header is supported by trimmer studs that pick up any loads above. This header—I usually use a doubled 2x10 with a 2x4 laid flat on the bottom—sits below the double top plates, which means the top of the window is about 13 in. below the ceiling.

With arch-top windows, however, it's common to want the top of the window closer to the ceiling, which means there's no room for a conventional header. Usually the arch top will stop 5 in. or 6 in. below the ceiling line, so that after installing the casing, you still have some clearance between the casing and the ceiling.

I use two methods to eliminate the conventional header. The first method is used if the window is in the first-floor wall of a two-story house. In this case, second-floor deck, wall and roof loads must be supported. I move the header up above the top plate (bottom drawing, facing page). In 2x4 framing, this would call for a double 2x10 with ½-in. plywood sandwiched between; in 2x6 framing, it would call for a triple 2x10 with ½-in plywood. In either case, I use the rim joist as the outside layer of the header. If the ends of floor joists bear above windows, I then install metal hangers to support the joists that meet at the header.

I use a different method if the window is in a wall supporting only ceiling and roof loads, where we often have less depth to work with. I usually have room for a smaller dimension header that's beefed up to carry the roof loads. For window openings 3 ft. to 4 ft. wide, I use a piece of ½-in. by 5½-in. steel sandwiched between 2x6s, or a piece of 4-in. by 4-in. steel angle (such as masons use for lintels) packed with two 2x4s. I bolt these headers together with ⅜-in. carriage bolts.

Framing and cathedral ceilings—On the house shown here (photo left), all the arch-top windows were in gable-end walls of rooms with cathedral ceilings. Cathedral ceilings present special problems because their construction usually requires a structural ridge beam. You cannot build a cathedral ceiling with conventional pairs of rafters opposing at a ridge board. Without ceiling joists or collar ties, the weight of the roof will push the outside walls apart. By making the ridge beam structural, however, the weight of the roof bears on the outside walls and on the ridge beam.

If there are no windows centered in the gable wall, the ridge beam is usually supported by a solid or built-up post, or by a header distributing loads to a pair of posts. When you place an arch-top window in the center of the gable, however, there is no room for the post and rarely is there room for a conventional header. Let's consider, for example, the 8-ft. dia. arch top I installed in the master bedroom. To support the ridge beam (a 5½-in. by 13-in. glulam) while keeping the arch-top window close to the peak, I turned the gable-end rafters into a truss.

First I made a pair of gable rafters out of 2x10s (the other rafters were 2x8s). I butted this first pair of gable rafters together in full plumb cuts. Then I fit a 2x12 crosspiece to the underside of the rafters so that the bottom edge of the crosspiece was even with the top of the rough opening for the arch top (top drawing, facing page).

Next I sheathed the inside face of this assembly with plywood, starting with a triangle whose sides followed the top edges of the rafters and the bottom edge of the crosspiece. I filled out the remaining faces of the rafters with ripped lengths of plywood.

Then I added another layer of 2x stock. But this time I ran a 2x12 chord all the way through—flush with the top edges of the rafters—and filled in with 2x10 stock above and below it. At the ridge, I notched the plumb cuts to create a pocket for the ridge beam. Next I added another layer of plywood, and finally, a third layer of 2x stock. I used construction adhesive between all the layers, and

Photo: Ross Cameron

From *Fine Homebuilding* magazine (October 1989) 56:64-67

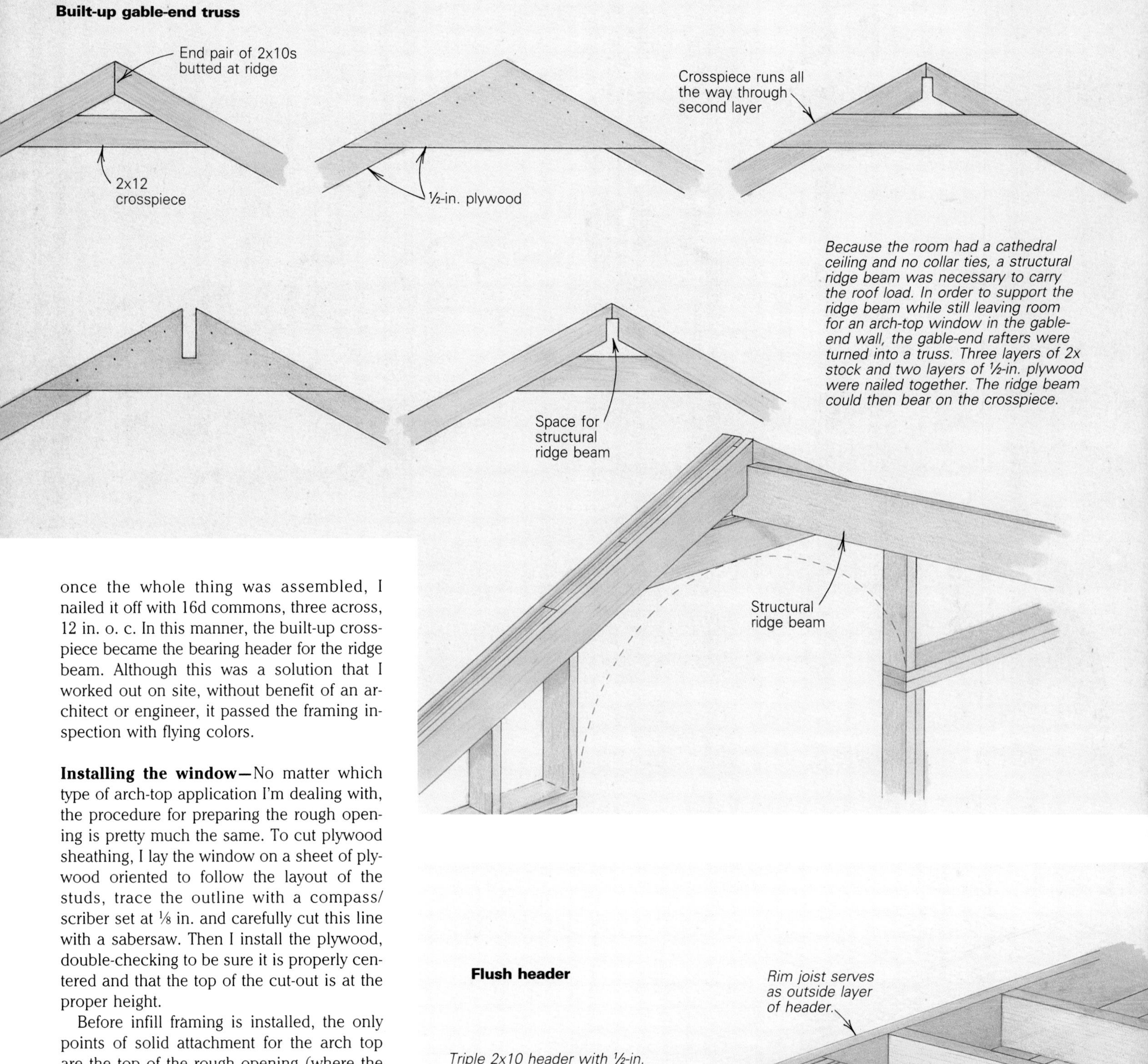

Because the room had a cathedral ceiling and no collar ties, a structural ridge beam was necessary to carry the roof load. In order to support the ridge beam while still leaving room for an arch-top window in the gable-end wall, the gable-end rafters were turned into a truss. Three layers of 2x stock and two layers of ½-in. plywood were nailed together. The ridge beam could then bear on the crosspiece.

once the whole thing was assembled, I nailed it off with 16d commons, three across, 12 in. o. c. In this manner, the built-up crosspiece became the bearing header for the ridge beam. Although this was a solution that I worked out on site, without benefit of an architect or engineer, it passed the framing inspection with flying colors.

Installing the window—No matter which type of arch-top application I'm dealing with, the procedure for preparing the rough opening is pretty much the same. To cut plywood sheathing, I lay the window on a sheet of plywood oriented to follow the layout of the studs, trace the outline with a compass/scriber set at ⅛ in. and carefully cut this line with a sabersaw. Then I install the plywood, double-checking to be sure it is properly centered and that the top of the cut-out is at the proper height.

Before infill framing is installed, the only points of solid attachment for the arch top are the top of the rough opening (where the arch is tangent to the bottom of the header), and at the two bottom corners of the arch top (where the arch is tangent to the gable studs). I install the window at this point, tacking it through the nailing flanges or exterior casing at these three points.

Then I fabricate a curved ladder of infill framing to provide nailing for the sheathing outside, and the drywall and casing inside (drawing, next page). I use 2-in. wide plywood for the curved "rails" of the ladder, then nail 5/4 by 3-in. blocks radially along the plywood every 6 in. to 8 in., except where I turn the blocks on edge so they don't stick out past the plumb-cut ends of the plywood.

I fit the ladders on either side of the rough opening, shim them snugly against the outside surface of the arch-top jamb and secure

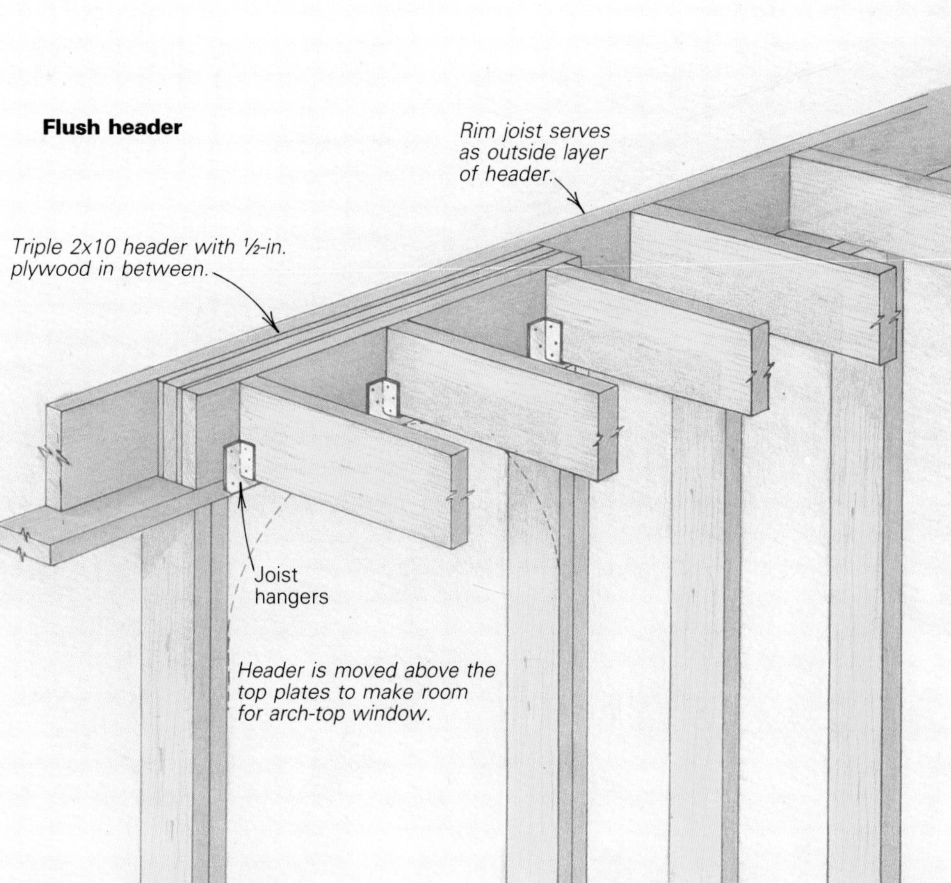

Header is moved above the top plates to make room for arch-top window.

Drawings: Michael Mandarano

To provide nailing for sheathing, drywall and trim around arch tops, Goodale installs curved ladders made of plywood sides with blocks nailed between them.

With the cedar siding tacked in place on the trammel jig, Goodale pivots the swing arm and neatly trims the boards for a perfect fit against the arch-top window.

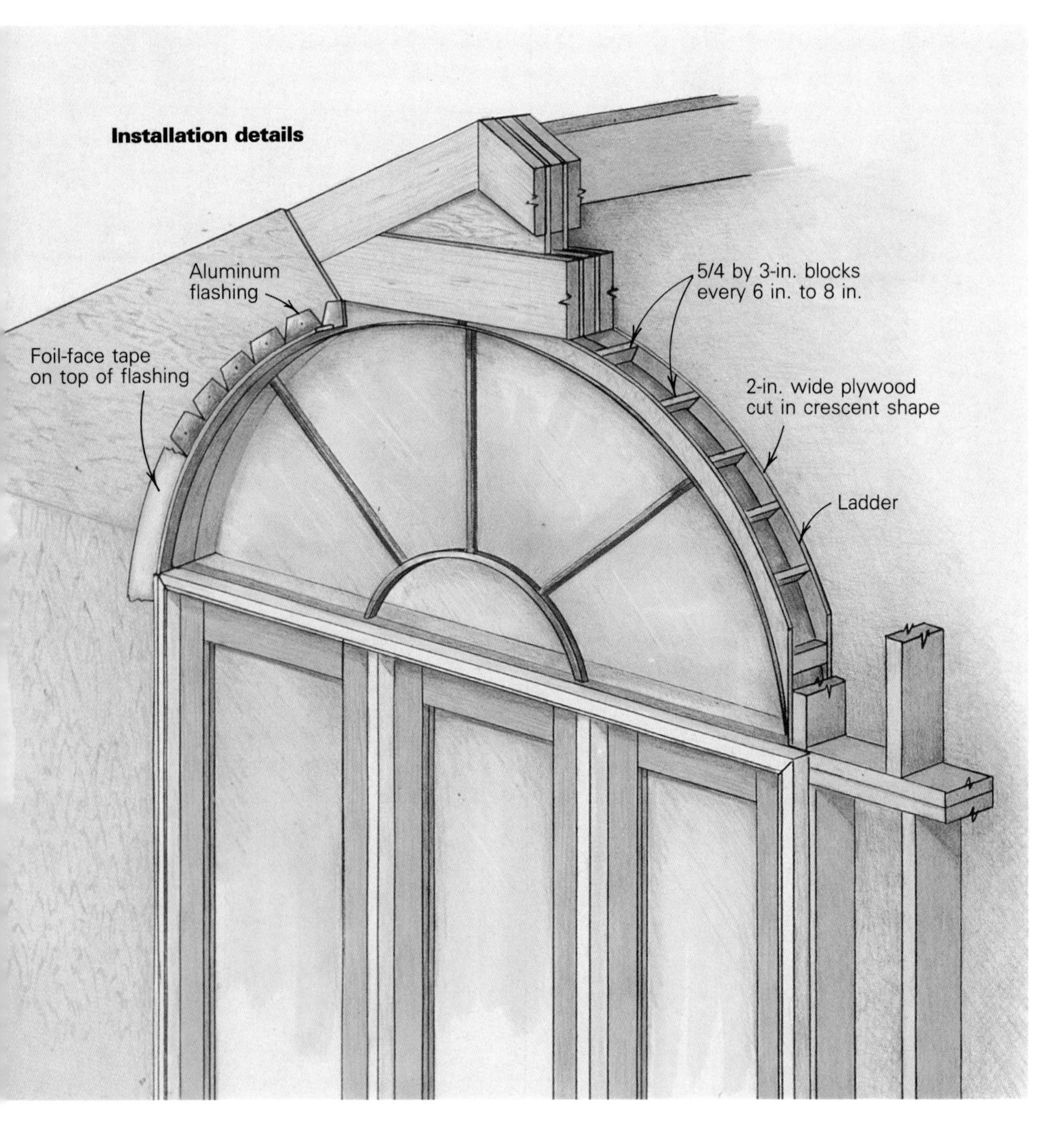

them in place by screwing through the sheathing into the ladders. They provide continuous nailing inside and out, and also allow me to insulate around the window right up to the jamb. Rather than bang nails into the lightweight ladders, I run #6 1¼-in. drywall screws through the nailing flange of the window and into the plywood.

Curved flashing—Most vinyl-clad windows are designed so that the nailing flange acts as flashing. Metal-clad and all-wood windows are more of a problem. Some manufacturers offer vinyl flashing as an accessory that you can buy. Or you can have custom flashings made (out of copper, for instance), but that gets pretty expensive. I usually make my own flashing on site. Sometimes I'll even add this to a vinyl-clad arch top so that its flashing will match the windows and doors on the rest of the house.

I use a sheet-metal brake to bend 6-in. aluminum flashing into a Z-shaped profile, the dimensions of which vary depending on the window. In this case the bends measured 4⅜ in. up the wall, 1⅛ in. across the top of the window and ½ in. over the edge of the window. Then I make cuts in the 4-in. edge, right up to the bend. The spacing of these cuts also varies with the radius of the arch top—the smaller the arch, the closer the cuts must be.

Another series of cuts about 2½ in. apart in the ½-in. flange allows the metal to bend to the shape of the arch top. When fitting the flashing, a helper supports the uncut end while I make the cuts in the ½-in. flange and fit it to the trim. The flashing is attached by nailing through the 4-in. flange

Photos this page: David Schiff

with roofing nails, using one nail per tab. I've tried cutting the flange on a bench, but it makes the stock too flexible to handle without distorting. As an alternative to cutting the ½-in. flange, you can crimp it with a three-leaf crimper (used by sheet-metal workers and woodstove installers).

I know that the cuts in the flashing are a vulnerable area, so I use a good quality foil-faced tape to seal the joint where the wall and window meet. I tear off short lengths of tape, and starting at the bottom, overlap the pieces as I work my way up the window. Lastly, I run a generous bead of clear caulk before installing siding. So far, none of my installations has leaked.

Using the trammel jig—My trammel jig is a worktable with an adjustable arm that swings arcs of any radius from 6 in. to 60 in. I use it as a giant compass to lay out half circles on plywood, and with a router mounted on the arm, I use it to cut siding and molding for arch tops.

To build the trammel jig, I start with a full sheet of plywood and add smaller sheets to the two bottom corners and the top edge of the jig. The corner panels are 16 in. by 24 in. The top panel is 16 in. by 48 in. These extension panels are attached to the full sheet with 1x3 cleats, glued and screwed through the back.

The adjustable arm is a length of ½-in. plywood, 4 in. wide by 60 in. long. It has a centered 1-in. hole 2 in. from one end and a ⅝-in. wide slot cut down the length, stopping 2 in. short of the 1-in. hole and 2 in. short of the other end. The router is centered over the 1-in. hole and screwed to the plywood.

The arm pivots on a carriage bolt running through the base of the jig and through the slot in the swing arm. The radius of any arc is determined by the position of the crosspiece, which has a hole drilled in it for the carriage bolt, and which can be screwed to the arm at any point along the slot. It's important that the pivoting end of the arm be shimmed up the thickness of the stock being cut.

On this house, the T&G cedar siding runs square to the pitch of the roof. The trammel jig allowed us to cut all of the pieces for the window at the same time so they fit perfectly. As a guide for the top edges of the stock, I tacked two boards to the table at the roof pitch. To use the jig, I tacked the stock to the trammel jig, carefully keeping the nails out of the line of cut.

With the stock positioned in the jig, I mounted a pencil in the swing arm and marked the line of cut on the stock. I removed the pieces and cut each one individually with a sabersaw to minimize the amount of work the router would have to do and to avoid blowing out the edges of the stock. Then I replaced the pieces on the jig and made the final cut with two or three passes of the router fitted with a ½-in. carbide straight cutter bit (top right photo, facing page).

A common location for arch-top windows is in the gable end of a room with cathedral ceilings, but this creates the problem of supporting a structural ridge beam. Goodale solved the problem with a site-built wood truss. The arched casing in the photo above was made on a trammel jig.

I also made interior casings with the trammel jig, beginning by mounting a pencil in the swing arm and drawing the inside and outside edges of the casing directly on the worktable. Then I layed out short sections of stock, overlapping each piece. The number of pieces needed to complete the arch varies depending upon the width of the stock—wider boards will require fewer pieces to complete an arc, but will also produce long run-out in face grain and should be avoided unless the casings are to be painted.

I joined the pieces by slot-cutting the matching ends and gluing them together with a spline. Once the pieces were glued up, I mounted the resulting polygon in the trammel jig, and using it as a compass again, redrew the inside and outside lines of the casing on the stock. Just as with the siding, I cut the stock close to the line with a sabersaw and then remounted it in the jig for final passes with the router.

To avoid putting unnecessary nail holes in the face of the casing, I held stock in the jig with drywall screws run through the back. Once the inside and outside edges were cut, I simply nailed it up (photo above). But you can switch the router bit to any desired molding profile and shape the edges or even the face of the casing. □

Douglas Goodale is a builder in Frenchtown, New Jersey. David Schiff is a writer and an amateur builder who assisted with the writing of this article.

Photo: Ross Cameron

Selecting Wood Windows

A look at what's in a manufactured wood window, how it operates and how it performs

by Joanne Kellar Bouknight

In a 52-acre factory beside the St. Croix River in Minnesota, I watch a load of rough 2-in.-thick wide pine boards drop in a single layer onto a 12-ft. wide conveyor belt, where each board is manually flipped to expose its worst side. A lumber grader sits in a swivel chair high over the planks, his hands ready on a keyboard covered with buttons. He eyeballs each board and manipulates laser lines on the board in a combination of established widths that will allow him to minimize defects and make the most economical use of the wood. He decides on the best combination and enters his selection on the board.

The grader's five-second operation is just the beginning of the route a piece of lumber takes at Andersen Corporation's plant in Bayport. Andersen, which sells more wood windows than any company in the world, is just one of many wood-window manufacturers enjoying the recent resurgence of interest in their products. Nowadays, builders, architects and homeowners building new houses are sorting through window catalogs from more manufacturers with more materials, colors, sizes, profiles and performance claims than they've ever seen before. This article surveys those windows, from the way they're put together to the way they're tested.

Looking back through windows—The first windows were simply wall openings providing observation, light or air. It wasn't long before oiled skins or paper and mica and marble sheets protected such openings, but glass was rare in windows of ordinary houses until the 1800s, even though 4th-century Romans (who enacted the first right-to-light laws) made windows from blown glass. Until the late 17th century, window glass was made by flattening a blown globe into a disk of up to 5 ft. in dia. or by cutting a cylinder that had been blown into a form. A Frenchman made the first flat glass in the 1600s, but flat-glass techniques didn't take off until the 20th century. Float glass—made by floating molten glass on molten tin, where it settles to a flat layer with parallel surfaces—is relatively new but it's now the standard for residential windows.

The two types of operating sash—sliding and swinging—aren't so new either. The early standard throughout Europe was (and still is) the casement window, at first with a fixed glazed sash above and a shuttered sash below for privacy and ventilation. Double-hung windows appeared in England in the 1600s and became the standard in the American colonies.

Sliding windows—Its association with Colonial styles makes the *double-hung* the favorite window type in the U. S., but it is also the least energy-efficient (along with its cousin, the horizontal slider). Because they must be free to slide, double-hung sash can't make an impregnable seal with the frame. Double- and single-hung windows are also harder to open than other operating types, making them questionable candidates for locations over sinks or for use by people with diminished strength. On the plus side, new double-hungs are easy to tilt in for cleaning. An advantage to a fully screened double-hung is that it can supply high and low ventilation at the same time. Even so, a double-hung is only 50% openable. *Single-hung* windows are cheaper and tighter than double-hungs, but can't provide top ventilation. Horizontal *sliders*, also called gliders, operate on the same principle as double-hungs (but they're easier to operate) and are endowed with similar advantages and disadvantages.

Swinging windows—Windows that swing open include casements, awnings, jalousies, hoppers, tilt/turns and an offshoot, the pivot window. Of all window types, except fixed, swinging sash offer the best protection against air infiltration because the weatherstripping is fully compressed on all four sides when the sash is closed. The casement cranks outward to a 100% vent opening and can deflect or capture the wind. When open, a casement sash stands a few inches away from the frame, allowing access to the outside of the glass for cleaning. Screens and sash are easy to install and the sash on today's casement window is easy to operate.

But an open casement offers no protection from rain—in fact, the hardware and inside of the sash are vulnerable if left open in the rain. In any weather the inside of a casement sash will get dirty faster than a sliding window. One builder I spoke with felt that it was tougher to keep a casement properly adjusted than a double-hung because of the more complex hardware. Its hardware and sturdy construction also make a casement a little more expensive than a double-hung of the same size and make. Because casements open out, it's important to provide a space between the fully opened sash and a walkway.

Awning windows are casements turned sideways but they can't deflect wind like casements and aren't quite as easy to clean. On the other hand, awnings and jalousies, which are multiple glass panels in the same frame, can provide both ventilation and rain protection at the same time. A *hopper*, which pivots inward, can provide rain protection if partly open in a gentle shower. Hoppers can interfere with blinds or furniture if they're placed low.

Tilt/turn windows are popular in Europe and gaining a foothold in the U. S. The tilt/turn's German-made dual-position hardware encircles the sash and allows a single handle to swing the sash inward about 5 in. at the top for ventilation and security or to swing it inward 180° at one side. When a tilt/turn is closed, sash bolts can lock into the frame at four to eight points around the window. Performance doesn't come cheap; a tilt/turn window costs more than a casement or double-hung and is often heavier. A *pivot* window sash swings from center points in the frame either vertically or horizontally. A big drawback to pivot windows is that the open sash is half inside and half out, making screening difficult.

Fixed windows are more secure and boast better thermal and air-infiltration rankings, but they don't provide ventilation and are impossible to clean from the inside unless you can reach the glass from an adjacent operable window. Fixed windows combined with ventilating windows will cost less than a bank of operable windows. And, because you don't have to screen fixed windows, you get more light year-round. Window companies offer an assortment of fixed-and-operable combinations for flat walls or as bay and bow windows. Geometrics (trapezoids and octagons, for example), roundtops and ellipses are also available to pair with conventional shapes.

Why wood?—The neighboring states of Minnesota, Iowa and Wisconsin are home to the oldest and largest window companies. White pine, the choice window material earlier in the century, was once plentiful here, while the Mississippi River and its tributaries offered ready transporation. Today the majority of wood windows are manufactured from Ponderosa pine logged in the Northwest. Ponderosa and other yellow and white pines commonly used in win-

From *Fine Homebuilding* magazine (April 1990) 60:46-51

dows are fine- and straight-grained, fairly uniform in color, and easily machined and glued.

Wood frames and sash boast good thermal resistance—at least 1,500 times better than nonthermally broken aluminum or steel and 70% better than all-vinyl. Though today's metal frames with thermal breaks rival the wood frame's thermal capacity, they are often more costly.

Weather is a wood window's nemesis. Frames and sash must be coated to resist the effects of rain and sun. As extra protection, frames and sash of clad and all-wood manufactured windows should be pressure-treated with waterproofing preservative before assembly. The waterproofing preservative's purpose is to retard swelling, shrinking and warping and to reduce attack by decay and stain organisms. In all-wood windows, preservative also helps paint to adhere. At Andersen, for example, sash and frame parts are pressure-treated with a non-Penta preservative for three minutes. The preservative penetrates ¼ in. perpendicular to the grain and 1½ in. at end grain. To circumvent an eight hour drying time at room temperature, treated wood remains in a drying chamber for about 15 minutes.

For best results the exterior of unprimed windows should be field-painted with an alkyd paint no longer than 4 to 6 weeks after installation. Most manufactured windows arriving on site are factory-primed on the exterior with latex paint. Interior priming is an option. Marvin Windows' standard is bare wood, with its finish options a latex primer or a high-performance coating called XL, which Marvin claims has a film integrity of 10 years. Other window companies offer similar coatings in 2 to 20 or more standard colors. Weatherstripping, hardware and jamb liners should not be painted.

For custom or very high-quality work that requires all-wood windows, you might consider frames and sash built of weather-resistant species like cypress, mahogany, or teak (see chart, pp. 60-61). Still, manufacturers of exotic-wood windows coat the wood with a stain or paint for a longer life and more uniform appearance.

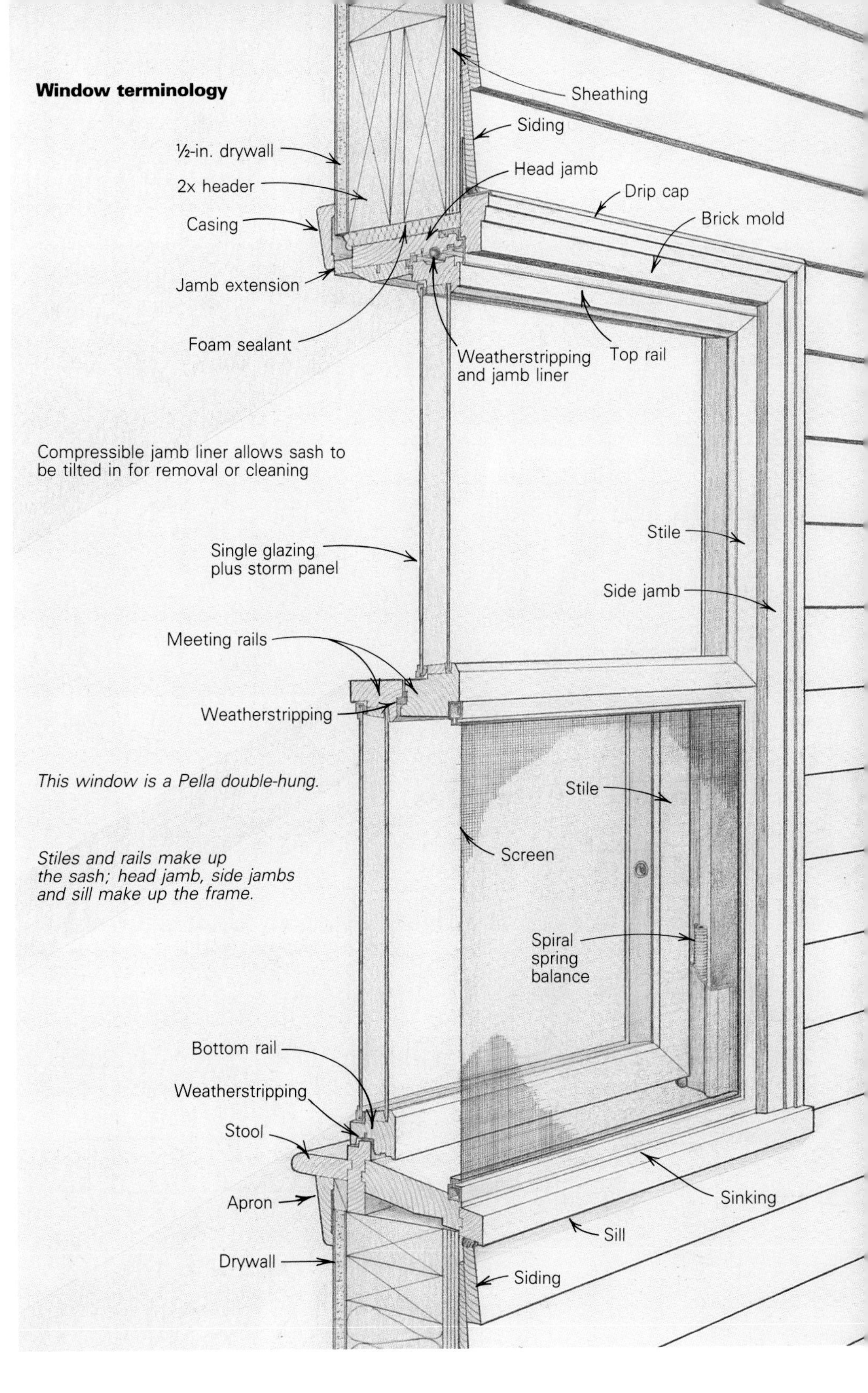

Aluminum and vinyl cladding—Cladding a wood window with aluminum or vinyl is another way to beat the weather and avoid repainting yet keep the look of a wood window inside. Many manufacturers make aluminum-clad wood windows, and a few make vinyl clad (see chart). Aluminum cladding, which is prefinished with an electrostatically applied coating, is durable, weather-resistant—and likely to make a window about 10% more expensive than its all-wood counterpart. Cladding on frames is generally .050-in. to .055-in. thick extruded aluminum. Sash cladding is .019-in. thick rolled aluminum, which is thinner in order to cut down on weight and bulk. Roll-form aluminum may also be used for frames. Pella clads its double-hungs entirely with roll-form and its casements with a combination of roll-form and extruded aluminum. When asked about the difference, Pella said that casement windows are more likely to be used in commercial work, where building codes require extra strength.

Extruded aluminum is designed to snap onto a wood subframe. It is screwed together and locked from the inside. Inside corners of some aluminum-clad windows may be pumped with urethane sealer though some companies feel this is unnecessary. Extruded aluminum may conform to the shape of the wood it clads, but it is often designed to pull away from the wood at vulnerable edges—such as the exterior edges of sill, head and jamb—to keep moisture from prolonged contact with the wood (drawings C and E, pp. 58 and 59). Water isn't expected to penetrate cladding under normal circumstances, but hollow aluminum edges will keep wind-driven moisture out of contact with the wood. (Roll-form aluminum cladding on sash or frame lacks rigidity, so it must conform to the shape of the wood.) True divided lights (TDL) on clad windows are sealed with a high-performance coating that looks like cladding but doesn't have the bulk. An exception is Pella, which covers the outside of its clad-window muntins with aluminum.

Aluminum cladding comes in standard colors: white, brown, and sometimes tan and grey. You'll pay a substantial setup fee for custom colors. Vinyl cladding isn't as widely used as aluminum, but Andersen has been working with the Italian-born process since 1962 and now makes all its stock windows that way. Some of

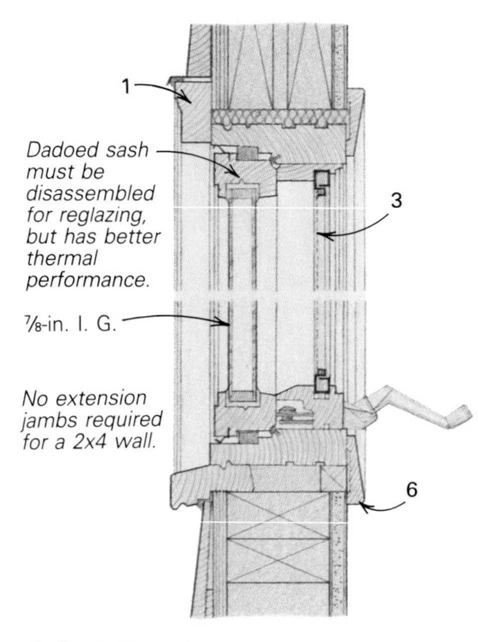

A. Pozzi all-wood casement in 2x4 wall with siding

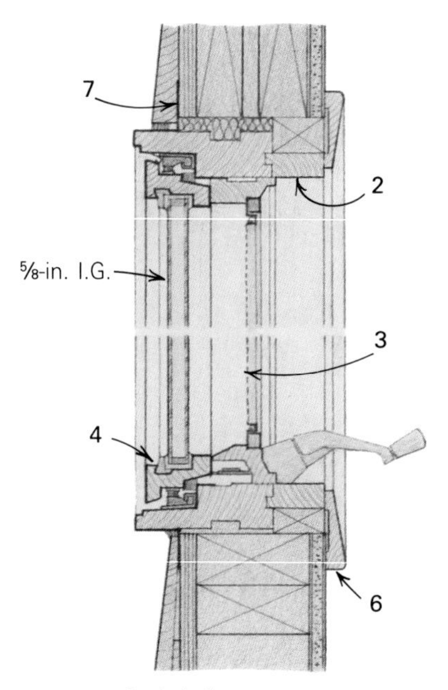

B. Andersen vinyl-clad casement in 2x4 wall with siding

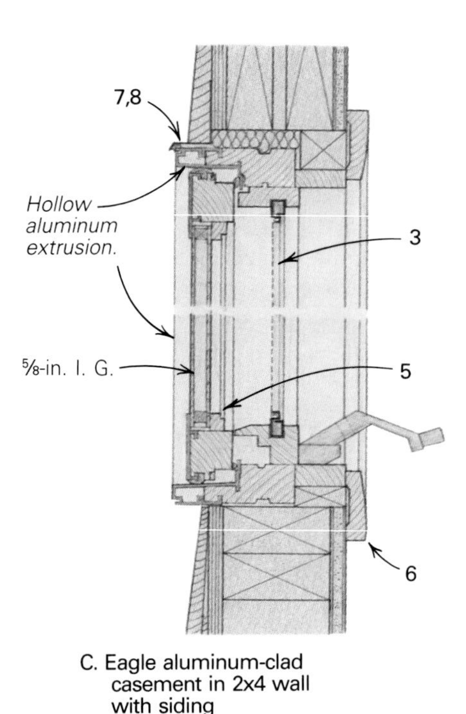

C. Eagle aluminum-clad casement in 2x4 wall with siding

Note: These details are taken from drawings in manufacturer's literature. All are drawn at the same scale. Sections are drawn through head jambs and sills. Some features shown may be optional. Glazing thicknesses shown are standard for each manufacturer. I. G. = insulating glass.

the components of a vinyl-clad window may not be clad, so the window may still need painting over the years. For example, an Andersen double-hung frame is clad, but its sash has a high-performance polymer coating similar to Marvin's XL (also with a 10-yr. warranty).

Because the properties of vinyl cladding limit the colors available, colors are limited to two: white and brown. Andersen allows that its brown—terratone—cladding can be repainted with a high-quality alkyd or latex paint, but white cladding, which differs chemically from brown, should not be painted.

Sash and frame details—As I toured Marvin's factory, I was surprised at how much of windowmaking remains handwork. Muntin bars are joined to sash, weatherstripping is applied and clad frames are snapped into place, all by hand. Joining the components into frames and sash is as mechanized as Marvin gets. But milling joints and profiles of sash, frame and muntins is the domain of the machines and high-speed molders. The milling operation is so precise that no sanding is required on these components.

In many wood frames—especially clad ones—components may be finger jointed. Finger-jointed and edge-glued stock can make frame components and structural mullions, but should only be used where the joints will be invisible and protected from the weather. Finger-jointed material can be stiffer, straighter and stronger than solid material, and the process is an economical use of too-short stock. The stiles and rails of manufactured wood windows are joined by slot and tenon or blind mortise-and-tenon joints and are secured with glue. The Architectural Woodwork Institute, which suggests standards to the architectural-woodwork industry, favors a glued and pinned mortise-and-tenon joint for premium and custom grades.

A frame is stationary, so it doesn't need the strength of a sash. Like a door frame, the double-hung window's head jamb is commonly housed in dadoes cut in the side jambs or joined to the side jambs by a double or single rabbet. Most casement frames are joined the same way, although you may find the stronger mortise-and-tenon joint at head and side jambs of a large window, such as Marvin's Magnum line.

Sash members range from 1⅜ in. to 2⅛ in. thick and 1³⁄₁₆ in. to 2⅛ in. wide. The standard jamb is 4⁹⁄₁₆ in., which fits with a minimum of planing into a 2x4 wall with ½-in. sheathing and ½-in. drywall. Aluminum and vinyl-clad frames are often slightly wider than 4⁹⁄₁₆ in. and will protrude from a 2x4 wall (drawings B,C,E and F, above). Clad windows don't need exterior casing, but if you want it for a more traditional look, you'll have to accept that protrusion. Jamb extensions, applied in the factory or on site, come in a wide range of sizes to allow the standard window to fit into any thickness wall (drawings above). A few window companies sell frames narrower than 4⁹⁄₁₆ in., in part to allow room for a drywall return.

A few window manufacturers make extra large, extra sturdy windows. Marvin's Magnum line, for example, features stock double-hung and hopper windows and custom tilt/turns (drawing F, above) with wider jambs, thicker sash members and beefier hardware.

Profiles of sash, frame and muntins give a window a distinctive look. The best way to look at stock windows is in a showroom or on a job site, but examining window catalogs will help give you a good idea of the differences, especially if you can compare details side by side. For example, although the bottom rail of a double-hung bottom sash is traditionally wider than other sash members (for both stability and proportion), only a few double-hung windows come with a wide bottom rail (drawings D and E above). And Tischler rounds all corners of its wood members to a 3-mm radius so that paint and stain will adhere longer.

Weatherstripping and hardware—A manufacturer can mill all its wood components or have someone else do it, but every manufacturer purchases weatherstripping, hardware, screens and, often, casings, snap-in grids and glazing from common vendors. That means you're likely to find strong similarities between such products, although they may be modified to match individual manufacturer's images and price tags.

Weatherstripping seals an operable window against air, water and dirt. Also important as the first defense against water is to slope the sill, provide sinkings or dams and form a drip in the bottom rail. Flexible, compressible materials work best as barriers against air, but sealing against water isn't quite so easy.

Swinging and pivoting windows are the easiest windows to seal. Neoprene and thermoplastic (vinyl) bulb weatherstrippings are the material of choice for compression seals on the frames. They are flexible and non-hardening in cold weather but have a high coefficient of friction (so they can't be used to seal side jambs of sliding windows). Leaf weatherstripping is applied to the outer edges of the sash, and bulb weatherstripping is applied to the face of the frame of the standard casement windows. Dual durometer vinyl bulbs have a hard fin attached to the wood and a soft, flexible bulb that compresses as the window closes. European windows, such as Tischler and Hobeka, use a neoprene bulb with vulcanized corners. Removable weatherstripping is a bonus for future maintenance or replacement.

Sliding windows are difficult to seal because many surfaces need protection. The wood jambs of today's double-hung windows are covered with a removable jamb liner of foam-lined vinyl. The foam compresses as the

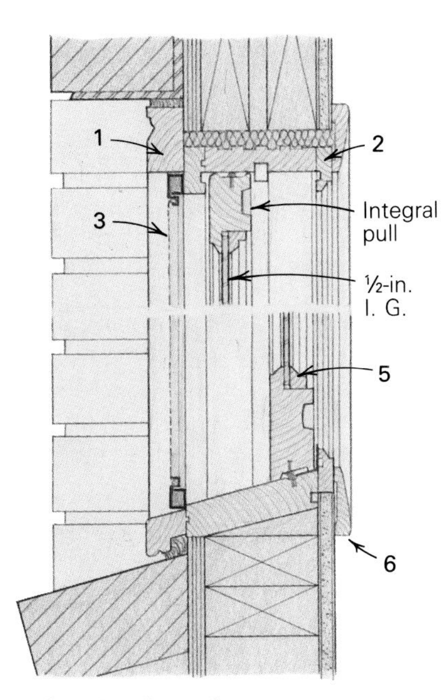

D. Kolbe & Kolbe all-wood double-hung in 2x4 wall with brick veneer

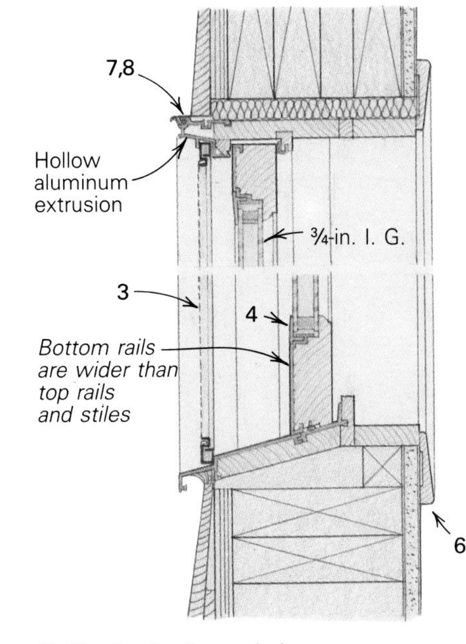

E. Marvin aluminum-clad double-hung in 2x6 wall with siding

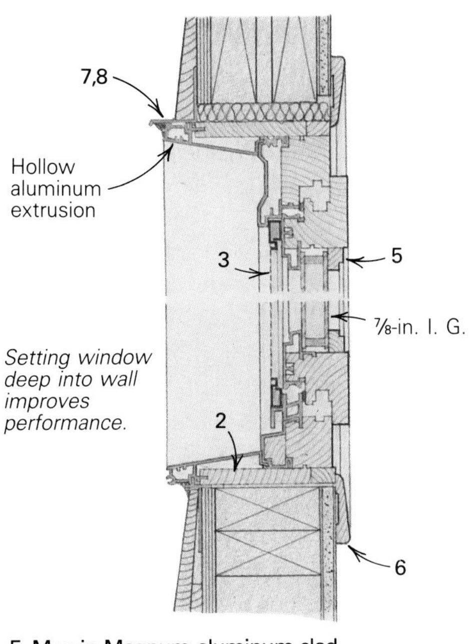

F. Marvin Magnum aluminum-clad tilt/turn in 2x4 wall with siding

Key
1. Brick mold
2. Extension jamb
3. Screen
4. Snap-in glazing bead
5. Wood stop
6. Picture-frame casing
7. Nailing flange
8. Drip cap

sash moves to provide a weather seal and allows the sash to be tilted into the room for cleaning or removal. PVC or polypropylene leaf weatherstripping is the standard for double-hung head jambs, while leaf or bulb weatherstripping is standard for the meeting rail and bulb weatherstripping for the bottom rail.

The standard finish color for window hardware is brown, bronze or gold; white is a common option. Corrosion-resistant hardware of stainless steel or nickel-and-copper alloy hardware is offered by a few companies (see chart).

A double-hung's hardware includes a sash lock (two on wide windows), a recessed or surface-mounted sash lift and balances—usually a coil-spring block-and-tackle system with nylon or dacron cording. A surface-mounted lift reduces the force needed to lift the sash.

Glazing—Window glazing has come a long way from oiled animal skins. Stock windows are available single-glazed or double-glazed; triple- and quad-glazing may be options. Note that a double-glazed window isn't always insulating glass (IG), which has a hermetically sealed air space. Double-glazing can also refer to single-glazing with a storm panel added, a system with a lower R-value and a greater tendency to exhibit condensation. Clear IG windows cost 10% to 15% more than non-insulating double-glazed windows. Today, the thermal performance of IG windows can be pumped up with low-e coatings and gas fills (see "High-Performance Glazing" in *FHB* #55, pp. 78-81). Hurd Millwork Company, for example, recently introduced what they claim to be an R-8 window.

Desiccant-filled aluminum spacers with double seals—most often polysulfide and butyl—are the standard for manufactured IG units. If you hate the look of shiny aluminum between the panes, ask if the spacers can be coated a different color (Pozzi's standard spacers are a neutral champagne color). But beware of dark colors, which will absorb heat.

Window manufacturers expect insulating glass to be free of fogging for 20 years or more, but as a rule most offer only 5-year warranties on glazing. You should be able to find the date the IG unit was made stamped on the spacer.

Standard glazing putty is still used in some manufactured single-glazed windows, and a silicone-base putty glazes all of Tischler's TDL windows. Standard glazing putty can degrade the seal of insulating glass, so most IG panels are held in place by a snap-in rigid gasket with a flexible vinyl tip, or by a molded wood stop. Snap-in glazing beads and wood stops are removable for reglazing.

If a catalog says that its sash is assembled with mortise-and-tenon joints and *screws*, that's a clue to look at the glazing detail. It could be that the glazing is housed in dadoes in the sash. That's the detail I found on Century, MW and Pozzi windows (except for Pozzi's TDL sash). The dado eliminates an assembly step, gives the sash a clean line and contributes to thermal performance, but it also means that you must disassemble the sash for reglazing, or order a new glazed sash.

True divided lights—Insulating glass is the TDL window's bugaboo. Muntin sizes in historic divided-light windows ranged from ⅝ in. wide on original Greek Revival windows to 1⅛ in. wide on Georgian and Colonial windows. Muntins that narrow weren't meant to support insulating glass of optimum width. Kolbe & Kolbe's 1⅛-in. bar is the narrowest TDL muntin I found, but it's available only on insulating glass with an overall width of ½-in.—narrower than the recommended outside width of ⅝ in. to 1 in. IG muntin bars 1⅜ in. to 1⅝ in. wide are the norm, but architects often don't like them that wide. As a compromise between a fat-muntin TDL and a removable grille with a thinner profile muntin, some window manufacturers offer narrow (⅞ in. or ¾ in.) muntin TDLs with single glazing, plus a one-light storm panel on the outside. Pella has a wide range of muntin sizes for true and false divided lights, including ⅞-in. muntins sliced in half and permanently bonded to single-glazed light. Pella's latest design is a ⅞-in. muntin sliced and bonded to both sides of one insulating glass panel with unsealed aluminum spacers between the muntin halves. Tischler offers the same design, and notes that a window with a single IG panel is easier to wire for security than a TDL window.

Ordering windows and accessories—You can get an idea of what a window manufacturer offers from its free catalogs, but if you order windows often, ask for an architectural detail manual with loose-leaf pages and ask about CADD window libraries. Detail manuals contain specifications, sizing charts and details at several scales. Pay attention to sizing charts, because conventions vary. Windows are traditionally designated by glass size—that's still the most common convention—but today's windows can also be specified by sash size, frame size or, rarely, rough-opening size. Eagle, Peachtree and Pozzi designate window units by the width and height of the manufactured frame. For example, C2040 means a casement that's 20 in. wide and 40 in. tall. This method is easy on designers, who look for frame size, and on builders, who can usually just add ½ in. to each dimension for rough-opening size. But even rough-opening requirements can vary: Marvin's rough-opening width is 1 in. wider than unit size, Pella's clad windows require ¾ in. extra to accommodate an applied subsill and Caradco requires only ¼ in. extra for height.

When specifying and pricing windows, keep a tab on the options. You may be expecting that windows automatically come with screens, but you'll likely have to order them as an extra. (For casement windows only, Pella still makes the rolling screen that launched its business in 1925.) Another accessory is the removable wood grille that attaches to the inside of an insulating or single-glazed unit or between panes in a single-plus-storm unit. Rather than purchase grilles from a vendor, Kolbe & Kolbe makes its own grilles to match the sash sticking.

The factory-installed blinds and pleated shades that window manufacturers offer can boost thermal performance, but not significantly. Uninsulated blinds or shades hung inside a window frame don't cut heat loss as effectively as insulated shades hung—or even better, sealed—against the frame. Roller and roman shades work better than blinds and 2-in. polystyrene boards stuffed into the window frame at night are best, but that's not something you're likely to try twice. Between-the-panes casement blinds won't help with privacy when the window is open. Vinyl-clad window manufacturers like Andersen warn in their catalogs that shading and insulating devices may cause thermal stress and/or deformation of the protective vinyl cladding.

How you plan to install the window will also affect ordering. Unless you're nailing an all-wood window into place through the jamb, you'll need nailing flanges, masonry clips and/or brick mold, or casing and a drip cap. Jamb extensions, interior head and jamb casing and sill stops are also options. Ask if stools are available because lumberyard stools require improvisation to fit around factory-installed jamb extensions, which are designed long to receive picture-frame casing. Rather than cut the stool to fit the jamb extension, some builders shorten the jamb extensions in place. If units will be mulled, or joined, together on site, include mulling accessories, such as structural mullions, caps and clips. A distributor will often mull units together before trucking them to the site.

A clad window can't be violated with a nail in the clad portion of its frame. Instead, clad windows come with masonry clips or nailing flanges, which can be installed in the factory or on site. Unflanged windows are easier to transport and store. On vinyl windows, nailing flanges can be formed integrally with the frame. A nailing flange reduces air and water infiltration around the frame, and an integral flange does that best. But if an integral flange breaks during construction, it could also breach the skin of the vinyl cladding.

Once the windows are ordered and any shop drawings are approved, you'll have a week to 15 weeks' wait. Distributors of windows often store stock sizes, so you could get small orders in a day or two. Custom windows from small, high-end or foreign companies take longer. Remember, though, that delivery claims aren't promises. Builder Charlie McLevy from Fairfield, Connecticut, told me that the problem he has with wood windows is getting service from the distributor when windows arrive with the wrong detailing or without the proper accessories—especially with custom orders, which the dealer doesn't like to take back. A window is one hole in the wall that costs a lot of money, he added, so if a distributor balks at correcting mistakes, take your problem to the manufacturer.

Company	Material	Operating types	Options/notes
American Woodwork Specialty Co. 4301 James H. McGee Blvd. Dayton, Ohio 45427 (513) 263-1053	Wood Redwood Vinyl clad	G, R	TDL
Andersen Corporation 100 4th Ave. N. Bayport, Minn. 55003 (612) 439-5150	Vinyl clad	A, BB, C, DH R, S	Salt-resistant hardware
Bilt-Best Windows 175 10th St. Ste. Genevieve, Mo. 63670 (314) 883-3571	Wood Aluminum clad	A, BB, C, DH R, S	TDL
Caradco Corp. P.O. Box 920, 201 Evans Dr. Rantoul, Il. 61866 (217) 893-4444	Wood Aluminum clad	A, C, DH, R, S	TDL
Century Windows 1301 Newark Rd. Mt. Vernon, Ohio 43050 (614) 397-2131	Aluminum clad with woodfiber and resin features	A, C, DH, G, R	TDL
Clawson Manufacturing Box 8891 Missoula, Mont. 59807 (406) 543-3161	Wood Aluminum clad	A, BB, C, DH, G, R, S	TDL
Crestline One Wausau Center 730 Third St., P.O. Box 8007 Wausau, Wis. 54402-8007 (715) 845-1161	Wood Aluminum clad	A, C, DH, S, R	TDL
Eagle Mfg. Co. 375 E. 9th St. Dubuque, Iowa 52001 (319) 556-2270	Wood Aluminum clad	A, BB, C, DH, G, R, S	20 special clad colors
Hobeka Openings Inc. 1060 Worcester Rd. Framingham, Mass. 01701 (508) 875-5514	Mahogany, spruce, oak, teak	C, G, R, TT	TDL All custom
Hurd Millwork Co 575 S. Whelan Ave. Medford, Wis. 54451 (715) 748-2011	Wood Aluminum clad	A, BB, C, DH S	TDL
KSI Building Products 26 MacArthur Ave. Cobleskill, N. Y. 12043 (518) 234-2561	Wood	BB	
Kolbe & Kolbe Millwork Co. 1323 S. 11th Ave. Wausau, Wis. 54401-5998 (715) 842-5666	Wood Aluminum clad (all roll-form)	A, BB, C, DH, G, S	TDL Magnum DH
Lincoln Wood Products Inc. 701 N. State St., P.O. Box 375 Merrill, Wis. 54452 (715) 536-2461	Wood Aluminum clad	C, DH, G	TDL
Louisiana-Pacific Corporation 324 Wooster Rd., North Barberton, Ohio 44203 (216) 745-1661	Wood Aluminum clad	A, BB, C, G, DH, R, S, SH	

Notes: Wood is a Western pine species unless otherwise noted.
Aluminum cladding extruded on frames and roll-formed on sash unless otherwise noted.

Window performance—"R-values really don't mean much." That's what I heard at the Minnesota Energy Council's conference, "Housing in Cold Climates" last year. During a two-day session on windows, the big issue discussed by manufacturers, scientists and builders was how windows are tested for thermal performance.

Today there is no national standard for testing and evaluating the thermal performance of windows. That means that comparing manufacturers' performance data is a little like comparing apples to lemons. The National Wood Window and Door Association (NWWDA) sets standards for air infiltration, water penetration, structural performance and operating force, but it doesn't look at thermal performance. And the six professional and trade associations that do set standards for and/or test windows and window parts don't work in tandem.

The two major sources of test methods for all types of windows are the American Architectural Manufacturers Association (AAMA), an association of steel-, vinyl-, and aluminum-window and door manufacturers, and the American Society for Testing and Materials (ASTM). AAMA and ASTM test methods differ, so comparing

Company	Material	Operating types	Options/notes
Malta Windows, Philips Ind., Inc. P.O. Box 397 Malta, Ohio 43758 (614) 962-3131	Wood Aluminum clad	A, BB, C, DH, S, SH	
Marvin Windows Warroad, Minn. 56763 (218) 386-1430	Wood Aluminum clad	A, BB, C, DH, G, S, P, R, SH, TT	TDL, curved glass, salt-res. hardware, 30 coating colors, square-cornered glass, Magnum DH & H
MW Manufacturers, Inc. P.O. Box 559 Rocky Mount, Va. 24151 (703) 483-0211	Wood Vinyl clad	A, BB, C, G, R	
NAWCO Minn. Inc. 1324 E. Oakwood Dr. Monticello, Minn. 55362 (612) 295-5305	Laminated wood Aluminum clad	A	Window also swings in for cleaning
New Morning Windows 10425-A Hampshire Ave. S. Bloomington, Minn. 55438 (612) 944-5120	Wood Aluminum clad	R Roundtop SH & TT	All custom TDL 5 clad profiles
Norco Windows, Inc. P.O. Box 140 Hawkins, Wis. 54530 (715) 585-6311	Wood Vinyl clad Aluminum clad	A, BB, C, DH, S	TDL
Peachtree Windows P.O. Box 5700 Norcross, Ga. 30091 (404) 497-2000	Aluminum clad	BB, C, DH, G	
Pella Rollscreen Company 102 Main St. Pella, Iowa 50219 (515) 628-1000	Wood Aluminum clad (rollform on DH frame)	A, BB, C, DG, G R, S, SH	TDL 20 optional clad colors, salt-res. hardware
Pozzi Wood Windows P.O. Box 5249 Bend, Ore. 97708 (800) 821-1016	Wood Aluminum clad	A, BB, C, DH, G, R	TDL
Tischler Und Sohn 51 Weaver St. Greenwich, Conn. 06830 (203) 622-8486	Mahogany, oak	TT & round P	TDL All custom
Weather Shield Manufacturing P.O. Box 309 Medford, Wis. 54451 (715) 748-2100	Wood Aluminum clad	A, BB, C, S	TDL
Webb Manufacturing Co. P.O. Box 707 Conneaut, Ohio 44030 (216)593-1151	Wood Urethane clad	G, R, and rotating R	TDL
Wenco Windows, Inc. P.O. Box 1329 3922 Lakeport Blvd. Klamath Falls, Ore. 97601 (503) 882-3451	Wood Aluminum clad	A, BB, C, DH, G, R	
J. Zeluck Inc. 5300 Kings Highway Brooklyn, N.Y. 11234 (718) 251-8060	Mahogany, teak, other specialty woods	A, BB, C, P, R, TT and triple-hung	TDL All custom

Key: A = Awning; BB = Bow and/or Bay; C = Casement; DH = Double-hung; G = Geometric; H = Hopper; P = Pivot; R = Round or roundtop; S = Slider; SH = Single-hung; TT = Tilt/turn; TDL = True divided light.

their results can be tough. Attendees of the Minnesota conference informally favored ASTM's test method, #C-286, and the National Association of Home Builders (NAHB) has set up a testing program following ASTM standards.

But after years of debate, window manufacturers, associations, government agencies and testing experts are finally getting together with the hope of setting uniform test standards. A uniform test will not only make it easier for builders and architects to compare manufacturers' product literature, but could lead to the development of an energy-performance label like the one on your water heater. To get an idea of what a uniform test might incorporate, I visited Chris Mathis, marketing director for Architectural Testing, Inc., in York, Pennsylvania, an independent lab that tests windows for NAHB, AAMA and individual window manufacturers. Mathis said to look for a national test method that will establish a "level playing field." Such a test should be performed by independent labs on random samples taken from production, and it should verify manufacturers' performance claims on a regular basis.

Until uniform testing standards are in place, anyone in the market for a window will have to read the fine print and ask questions. If you don't find values for thermal performance qualified in the catalog, ask the manufacturer if its values are obtained by testing or by calculation, if the results reflect one window or the average of many, and if values are for glazing alone or for the whole window.

Testing windows and averaging the results will give more credible results, but calculating thermal performance is common. To calculate the thermal resistance of a window, a manufacturer "corrects" resistance values for glazing with values for frame types. Values come from the *Handbook of Fundamentals*, published and revised every four years by the American Society of Heating, Refrigeration and Air-Conditioning Engineers (ASHRAE). This method was reasonably accurate until the advent of high-performance glazing systems, which haven't been thoroughly tested until recently and which often have better thermal resistance than frames. That means manufacturers may skip the frame and advertise values for glazing alone. But ASHRAE's 1989 handbook has tightened up its resistance values and takes a closer, tougher look at frames.

Calculated thermal-performance values that aren't verified by testing are always open to question, but once the 1989 ASHRAE values begin appearing in product literature, you can be more confident of calculations, especially if both glazing and frame are taken into account. Right now, though, don't be too picky about fractions: the Minnesota Energy Council and the Lawrence Berkeley Laboratory in California found that 60% of installed windows tested for thermal performance didn't reach advertised values.

Air infiltration can make a bigger dent in a heating bill than thermal performance, so it's also wise to pay attention to air infiltration rates. All wood-window manufacturers listed in the chart exceed NWWDA's air-infiltration standard by miles, so don't be impressed with statements to that effect. Inquire instead if results were obtained from a single window or from a larger sample, which will be closer to reality.

Window manufacturers at the cold climate conference also advised consumers to ask manufacturers: What's your service policy for manufacturing defects? How long have you made this detail or type of window? How do you test materials and designs for long-term performance?

Another way to be sure you're getting the best thermal performance is to take a close look at window details and placement. For example, setting the glazing deep in the sash improves energy performance. And although certain combinations of temperature and relative humidity will always cause condensation, using insulating glass instead of single-plus-storm will lower the incidence of condensation. So will placing the window close to the interior to provide a protected air space on the outside and to allow warm air to flow across the glass on the inside. It helps to locate supply registers directly under or above a window, too. Condensation is worse on larger panes of glass, so double-hungs or TDLs may be preferable to one-light casements of the same frame size. □

Joanne Kellar Bouknight is an assistant editor of Fine Homebuilding.

Simple Joinery for Custom Windows

Time-saving lap joints can save money, too

by David Frane

Few builders enjoy midstream design changes, but most of us have learned to deal with them. There are times, though, when a small change can cause a big problem. An example of this came while we were building a large home outside of Boston, Massachusetts. The roofer had finished about half of the red cedar shingle roof when the architects literally threw us a curve: The client had decided to dress up the front facade with an eyebrow window installed high above the main entry (photo right).

This design change created two problems. First, our window supplier wanted too much money (about $3,000) to custom-build the unit; second, they couldn't deliver it for three months. We didn't want to frame the dormer without the window—the fits would be too exacting and mistakes too expensive to fix—so we decided to build the eyebrow ourselves.

After studying the architects' rough elevations, I told job supervisor Harry Irwin that we could build the unit for about half the quoted price—*if* the client would accept some unconventional and slightly archaic construction details (we actually did it for about one-quarter of the original quote). And we could have the window ready to install in 10 days. The architects and the client, pleased that we'd found an affordable alternative, quickly gave us the go-ahead.

Looking out from an empty attic, the unit (top drawing, facing page) would feature a four-lite hopper (a bottom-hinged window that opens inward). On either side would be two fixed sashes, and the entire assembly would be surmounted by a curved top jamb screwed to a sill beveled to improve drainage. A curved casing applied to the top jamb would serve as a crown molding over which the roof shingles would extend. The operable window was overkill; the owners previously had lived in old houses with hot attics, so they wanted to have plenty of ventilation up there this time. As it turned out, the hopper is rarely opened.

My plan was to build the individual window sashes first, screw them to the sill and then laminate the jamb around them. I could then apply the window stop and casing and haul the unglazed unit to the roof to be framed into place. After the roof was finished, someone would have to go back up and glaze the sash.

An unexpected curve. **Lap joinery, used by the author to build this eyebrow window, offers an alternative to cope-and-stick joinery. Photo by Charles Wardell.**

Simplified joinery—To save time, we decided to use old-fashioned glazing compound instead of curved wooden stops to hold the glass in place. The window would be 35 ft. off the ground, so nobody would notice the substitution. And rather than using standard cope-and-stick joinery, I designed the sashes to incorporate lap joints.

Using lap joints probably saved a lot of time. Also, the multiple shaper setups and the jigging required for coping and sticking curved stock would have busted our budget. Though lap joinery is unconventional for a window, it's as strong as the conventional method. Besides, from the ground it would be indistinguishable from the joinery on the house's other windows.

Laying out the curves—Lap joints notwithstanding, making an eyebrow window remains technically demanding. The unit must be symmetrical and its curves fair—they must flow smoothly into one another with no visually jarring transitions. The overall shape of this unit would be formed by three tangentially intersecting arcs having identical radii (bottom left drawing, facing page). Two concave curves—the ends of the window—would meet the ends of a single convex curve—the window's center arc. The window would measure 12 ft. end to end and 2 ft. from the sill to the apex of the center arc.

We began by drawing a full-scale pattern of the window's perimeter (representing the entire unit with the casing installed) on ¼-in. lauan plywood. Jim Garry, a member of my crew who has lots of shop experience, laid out the curves with a long trammel. It took some work to find a combination of centerpoint locations that produced the right shape with the proper dimensions.

Next came my turn. I got the profiles of the three sashes by drawing a series of lines parallel to the inside of Garry's pattern. These lines represented the lower edges of the casing, head jamb and window stop, as well as the curved top

From *Fine Homebuilding* magazine (August 1992) 76:72-75

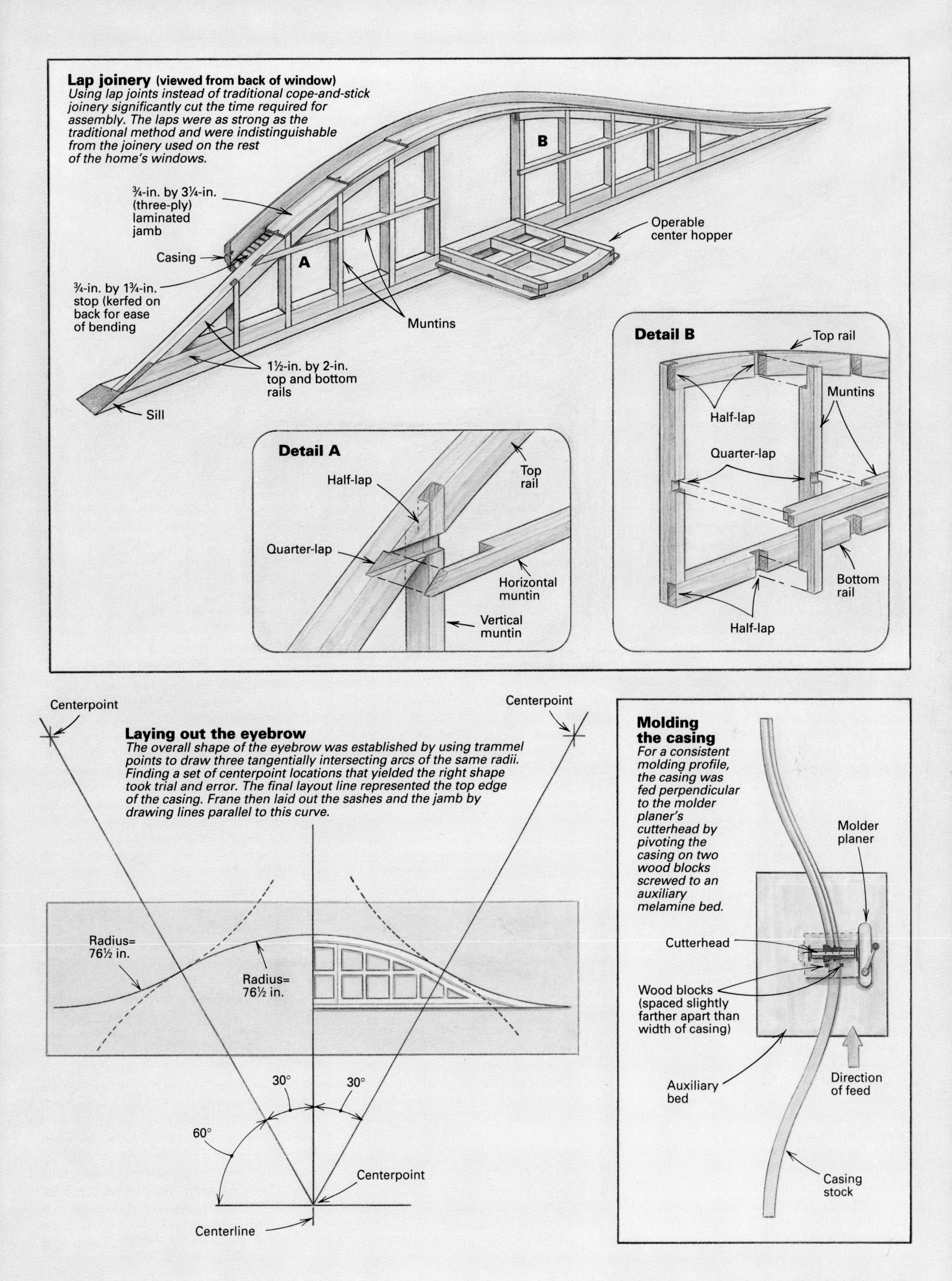

Drawings: Bob Goodfellow

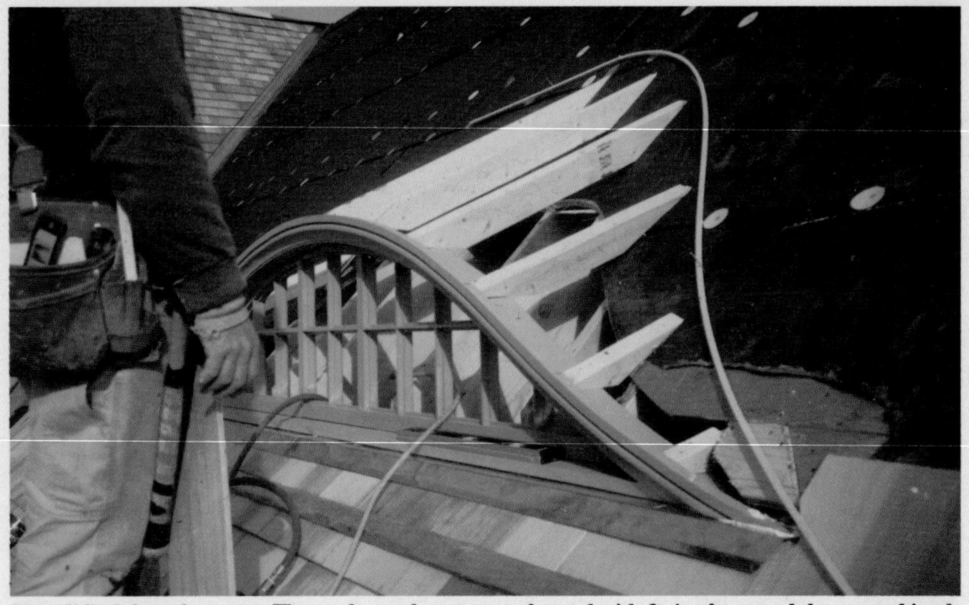

Simplified framing, too. **The eyebrow dormer was framed with 2x4 rafters, and the curved jamb was used as a header. Because the window looked out from an unfinished attic, the roof rafters behind it could be left intact. A pine batten is being used here to lay out a fair curve on the main roof.**

Section through hopper

Cedar shingles over elastic sheet membrane

Curved casing

2x4 eyebrow rafter

Laminated jamb

Stop

Hopper

Hinge

Hopper opens between two roof rafters

Sill

Stop

Copper flashing

2x12 roof rafter

rails of the sashes. I generated each line by making a series of pencil marks the proper distance down from Garry's curve, bending a thin wooden batten so that one of its edges touched all the marks, then connecting the marks with a pencil line scribed along the batten. At the bottom of the pattern, I drew straight lines for the sill and the lower sash rails, then drew in the muntins.

Making curved sashes—The first step in making the curved top rails of the three sashes was to transfer our curves from the lauan pattern to the solid white-pine sash stock. To do this, I borrowed an old boat-building trick (drawing below). I laid 5d box nails flat on the pattern stock with their heads on the curved line I wanted to transfer. I tapped the nail heads into the lauan so that they wouldn't move, then carefully lowered the sash blanks onto the pattern and tapped down on them to make an imprint of the nail heads on the blank. Then I drew lines through these marks, roughed out the blanks on the bandsaw and smoothed them with a spokeshave.

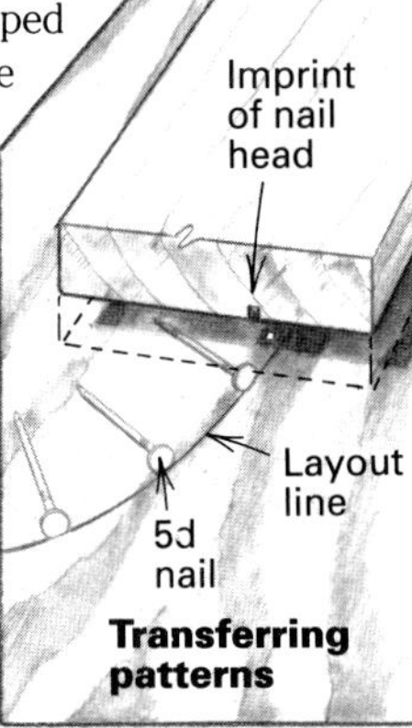

Assembling the blanks into sash frames was straightforward. I simply laid the blanks on the pattern, ticked off the intersections of the stiles and the rails on the sashes and cut my half-laps with a dado head on the radial-arm saw. On curving cuts, I roughed out the rabbets with the dado head and finished them with a chisel. Routing would have been more precise, but it also would have required more setup time and would have given me a better job than I needed in this case.

I glued all the pieces together with an epoxy adhesive. For this job I used Chem-Tech's T-88 (P. O. Box 70148, Seattle, Wash. 98107; 206-783-2243). This product is as thick as honey and harder to use than many other epoxies. But we were working in an unheated shop and T-88 is the only epoxy I know of that will set at temperatures down to 35° F.

I glued the half-lapped stiles and rails together on top of the full-size pattern. Clear plastic sheeting placed between the sash and the pattern kept the two from being glued together. The sashes were to be paint grade, so instead of clamping the joints, I simply drove long drywall screws with washers through the joints and the lauan pattern and into the underlying wooden table.

When the epoxy set, I removed the screws and cleaned up the excess epoxy (the screw holes would be puttied later). Then I transferred the muntin locations from the pattern to the new sash frames. My preferred tool for dadoing the half-laps would have been the radial-arm saw, but our saw didn't have enough throw to reach across the sashes. Instead, I made multiple cuts with a Porter-Cable trim saw, then cleaned out the dadoes with a chisel. Dropping the vertical muntin bar stock into these dadoes, I then marked the locations where the horizontal muntins would intersect vertical muntins, stiles and rails. Because the horizontal muntin at each end of the window would intersect a half-lap on the top rail (detail A, previous page), I decided to install all of the horizontal muntins using quarter-laps instead of half-laps. I cut these quarter-laps by running the entire sash frame through the table saw, keeping the bottom of the sill against the rip fence. After testing for fit, I glued the muntins to the sash frames. Using a rabbeting bit in a laminate trimmer, I then cut rabbets on the exterior of the window to accept the glazing and the putty. The rabbets were squared with a chisel. On the interior, I used the router to cut an ovolo profile around each lite opening (hardly anyone will ever see this, but it made me feel better). Now it was time to make the sill and the jamb.

Beveled sill, curved jamb—Making the sill and the jamb was the simplest part of the job. The sill is a single piece of 1½-in. thick pine ripped and beveled on the table saw. Because the sill would butt against the head jamb at both ends, I marked the end cuts of the sill by tacking the sashes to the sill and projecting the curvature of the jamb across the edges of the sill. Then I removed the sashes, cut the sill on the bandsaw, reattached the sashes and turned my attention to the jamb.

I laminated the jamb out of three 4-in. wide by ¼-in. thick strips of pine. Instead of building a separate laminating form, though, I used the tops of the sashes themselves. With the help of Don Pascucci (who would frame the unit onto the roof), I nailed the first lamination to the tops of the fixed sashes, then epoxied and nailed the successive laminations over it. After the glue set, I planed the rough edges of the jamb with a portable power plane, then trimmed its ends flush with the bottom of the sill.

Finally, I nailed a curved 1x1 pine stop to the top jamb. Because the stop would show a mere ¼-in. reveal beneath the top casing, I bent the stop by cutting a series of cross-grain 7⁄16-in. deep kerfs in its backside. Epoxy secured it to the jamb, and putty filled the kerfs.

Milling reverse curves—Garry made the casing from 5/4 stock. After transferring the curves from the pattern to the stock, he cut his pieces and joined them into a single casing blank using long scarf joints and epoxy. The pitch of the scarf must be 1-in-12 or less; otherwise you're just gluing end grain, and the joint won't hold.

Of course, the main challenge posed by the casing was that, unlike the rest of the window, there was no way to escape having to mold a profile on it. We had run plenty of curved stock

Where to buy eyebrow windows

The following companies offer either stock or custom windows that can be used for eyebrows. This is not a complete list, however. You should also check with your local suppliers or contact the following trade associations: The National Wood Window and Door Association (708-299-5200) consists of window and door manufacturers. The National Sash and Door Jobbers Association (708-299-3400) consists of window and door distributors.

***— Mark Feirer, editor of* Fine Homebuilding.**

Andersen Windows, Inc.
100 Fourth Ave., North Bayport,
Minn. 55003-1096
(800) 426-4261
Low-e, argon-filled round-top and elliptical units. Wood or vinyl-clad. Extension jambs available. Laminated and curved trim available.

Atrium Door & Window Co.
P. O. Box 226957, Dallas, Texas 75222-6957
(800) 527-5249
Half-rounds, prefinished (primed line will soon be available), true divided lite or snap-in grills. High-performance glazing (HPG).

Caradco
P. O. Box 920, Rantoul, Ill. 61866
(217) 893-4444
Half-round, elliptical and quarter-round windows. Custom windows also available. Aluminum-clad or primed wood exteriors. Natural wood interiors. HPG.

Crestline
One Wausau Center, P. O. Box 8007,
Wausau, Wis. 54402-8007
(800) 552-4111
Stock and custom. HPG. Wood and clad.

DashWood Industries, Ltd.
Box 10, Centralia, Ont., Canada N0M 1K0
(519) 228-6624
Aluminum-clad, vinyl-clad, encapsulated. Bare wood and primed.

DF Windows
Donat Flamand, Inc., 90, Industrielle St.,
Saint-Apollinaire, Que., Canada G0S 2E0
(418) 881-3974
Wood round-tops and half-rounds.

Hurd Millwork Co., Inc.
575 S. Whelen Ave., Medford, Wis. 54451
(800) 2BE-HURD
Standard sizes in half-round, quarter-round and ellipse. Custom sizes and shapes available (aluminum-clad or primed wood). HPG.

JJJ Specialty Co.
113 27th Ave., N. E., Minneapolis, Minn. 55418
(612) 788-9688 or (800) 445-6736
Wood and aluminum-clad windows. Elliptical and round-top windows. True divided lite or single lite with grill. Custom sizes only.

Kolbe & Kolbe Millwork Co., Inc.
1323 South 11th Ave., Wausau, Wis. 54401
(715) 842-5666
Round-top, half-round windows. Laminated jambs. Stock and custom. True divided lite and single lite with grill. Any wood species available in custom line.

Lincoln Wood Products, Inc.
P. O. Box 375, Merrill, Wis. 54452-0375
(715) 536-2461
All units are custom. Aluminum-clad (four colors) or unfinished. Round-top or quarter-round. HPG. Tinted or tempered glazing available.

Loewen Windows
1397 Barclay Blvd., Buffalo Grove, Ill. 60089
(800) 245-2295 or (708) 215-8200
Stock and custom. Wood and clad. HPG.

Marvin Windows
P. O. Box 100, Warroad, Minn. 56763
(800) 346-5128
Custom and standard round-tops. Aluminum-clad or unfinished. True divided lites or single lite with grill. Prefinished in standard colors.

New Morning Windows, Inc.
11921 Portland Ave. South, Burnsville,
Minn. 55337
(612) 895-6175
Custom with any glazing. Wood and clad.

Norco Windows, Inc.
P. O. Box 140, 811 Factory St., Hawkins,
Wis. 54530-0140
(800) 526-3532 or (800) 826-6793 (Wis.)
Half-round, custom and stock windows. Unfinished or primed wood exterior.

Peachtree Doors and Windows, Inc.
Box 5700, Norcross, Ga. 30091
(800) 477-6544
Custom. Aluminum or wood. HPG.

Pella Windows & Doors/Rolscreen Co.
102 Main St., Pella, Iowa 50219
(515) 628-1000 or (800) 524-3700
Stock round-tops. Aluminum-clad and finished or unfinished wood. Standard and custom colors.

Pozzi Window Co.
P. O. Box 5249, Bend, Ore. 97708
(800) 821-1016
Custom and stock half-rounds. Wood (single lite or true divided lites) or wood-clad. HPG.

Wenco
P. O. Box 259, W. Main St., Ringtown,
Pa. 17967
(800) 255-7743
Stock and custom units. Aluminum-clad or unfinished wood. HPG and tempered glass available.

Zeluck, Inc.
5300 Kings Highway, Brooklyn, N. Y. 11234
(718) 251-8060, ext. 89
Custom windows only. Various woods, including mahogany, teak, walnut and cherry. HPG.

through our Williams & Hussey molder/planer in the past, but this casing was different. Making molding requires that the stock be fed straight into the cutterhead. With simple, curved pieces, the usual technique is to register the stock against curved guides. But the curves on the eyebrow casing reversed direction twice. Garry's low-tech solution was to free-hand the blank through the molder/planer by pivoting it on two wood blocks that he has screwed to an auxiliary melamine bed (bottom right drawing, p. 63). Garry positioned the blocks on the infeed side of the cutterhead, letting him use the blocks as he would the guide pins on a shaper table. The result wasn't furniture grade, but it was up to snuff as exterior architectural millwork. The casing was applied to the window, everything was sanded and primed, and the hardware was affixed to the operating sash. We were ready for installation.

Framing the dormer—At this point, Pascucci took over. Because the window would look out from an empty attic, the only headered opening needed was behind the operable hopper sash; full rafters would run behind the rest of the unit (for more on framing a full eyebrow dormer, see *FHB* #65, pp. 80-84). All the rafters above the window were painted black so that they wouldn't be visible from the street. Pascucci's framing technique let him support the window on seat cuts made in the top edge of the existing rafters (top drawing, facing page). Using a pine batten, Pascucci then drew in a fair curve corresponding to the intersection of the eyebrow's rafters with the main roof deck (photo facing page).

The dormer framing consisted of short 2x4 rafters that were screwed to the window's head jamb at one end and to the main roof at the other end. Sheathing this was a bit tricky. Because the ¾-in. plywood we used elsewhere wouldn't make the necessary bends, Pascucci used three layers of ¼-in. lauan instead. The lauan was applied in overlapping strips because full sheets could not easily conform to a compound curve. Fortunately, our roofer was able to blend the eyebrow dormer's cedar shingles into those of the main roof. The shingles were applied over an elastic-sheet membrane to keep any water that backed up under the shingles from leaking into the attic. At the junction of the two roofs, copper step flashing was hidden between courses in what was, in essence, a woven valley. This meant carefully choosing pliable shingles and, when that wasn't enough, boiling them. □

David Frane is a foreman with Thoughtforms Corp., a construction company in W. Acton, Mass. Photos by the author except where noted.

Double-Hungs Restrung

More than just fishing for sash weights

by David Strawderman

I get a lot of calls from clients who want me to replace their windows. Often the old windows are wood sash that have been neglected and are on the verge of falling apart. And sometimes I'm hired to yank out a wall of replacement metal windows that clash with the style of the house. There are many instances, however, when it's best to restore the original windows to working order. This was certainly the case with John and Nina Heaths' recently acquired home. Located in the elegant Hancock Park district of Los Angeles, the house was originally constructed in 1923 by the contractor who built many of the fine surrounding dwellings.

The house was structurally sound, but it had suffered many remodels and lost a good deal of its original refinement. And a lot of things didn't work—like 30 double-hung windows.

Windows with a twist—The windows appeared to be regular old-fashioned double-hung sash of the type popular from the turn of the century to the second World War (drawing below). They were equipped with cords, pulleys and iron weights for counterbalance. A traditional thumb-latch secured the upper and lower sash and a recessed brass finger-pull assisted in raising the lower sash. But unlike other double-hung windows that I had worked on, this finger-pull had a small knob in its center, the significance of which I would soon learn. A closer inspection revealed nice touches. All the interior stops were secured by screws rather than by the usual finish nails. The lefthand parting strips were affixed with flathead screws, a clue that dismantling should proceed from that side.

Most of the windows were painted and caulked shut, but I found one that could still be opened. When I lifted the lower sash, I was surprised to find a split sill containing a recessed wooden bar. A spherical-headed screw protruded from the bar, and when I pulled on it with a pair of pliers the wooden bar revealed itself as the top rail of a concealed window screen. I now understood the function of that knob in the finger-pull: it operated a claw mechanism that clasped the screw protruding from the screen frame. A 90° turn released the screw so the sash could be opened without the screen. The upper sash had a mechanism with a similar device. I was working on a houseful of windows with retractable screens.

It took me two hours to partially dismantle the first window. As it turned out, nearly one man-day was needed to return each window to full working order. Although the retractable screens presented special problems, the steps I took to rejuvenate the windows are similar to those required for all double-hung windows.

Curing sash paralysis—The first step in restoring the windows to full operation was to free each sash. I used a sharp utility knife to cut the paint and caulk between the sash and the stops. This reduces splintering and paint chipping along the intersections of the stops and the sash. Next I used a small flat bar to gently pry the corners of the sash away from the stops. It's a bad idea to pry from the center because old sash rails are often weak, and excessive pressure will crack the glass. For removal, the lower sash must lift enough to clear the sill lip, and the upper sash must move down enough to clear the pulley assembly.

Next I removed the left interior stop. Because these were secured with screws, they were relatively easy to dismantle. Most interior stops, however, are fastened with 6d finish nails, and they should be pried loose with a wide putty knife. I begin near the center of the stop, as the ends tend to be encrusted with thicker layers of paint where they abut the upper stop and sill. Once the middle is free, the ends usually pop out. Then I remove the nails and set the stop aside.

The lower sash of a typical double-hung window can be maneuvered out of its channel once a stop is removed. But these weren't typical windows. They had flashings along their stiles that interlocked with metal channels affixed to the parting strips. As a consequence, I also had to remove the left parting strip before I could pull out the lower sash.

Parting strips are usually unpainted, making their extraction a pretty straightforward process. The parting strip is a single ⅜-in. by ⅝-in. piece of wood let into a groove in the jamb. It is usually secured with three or four

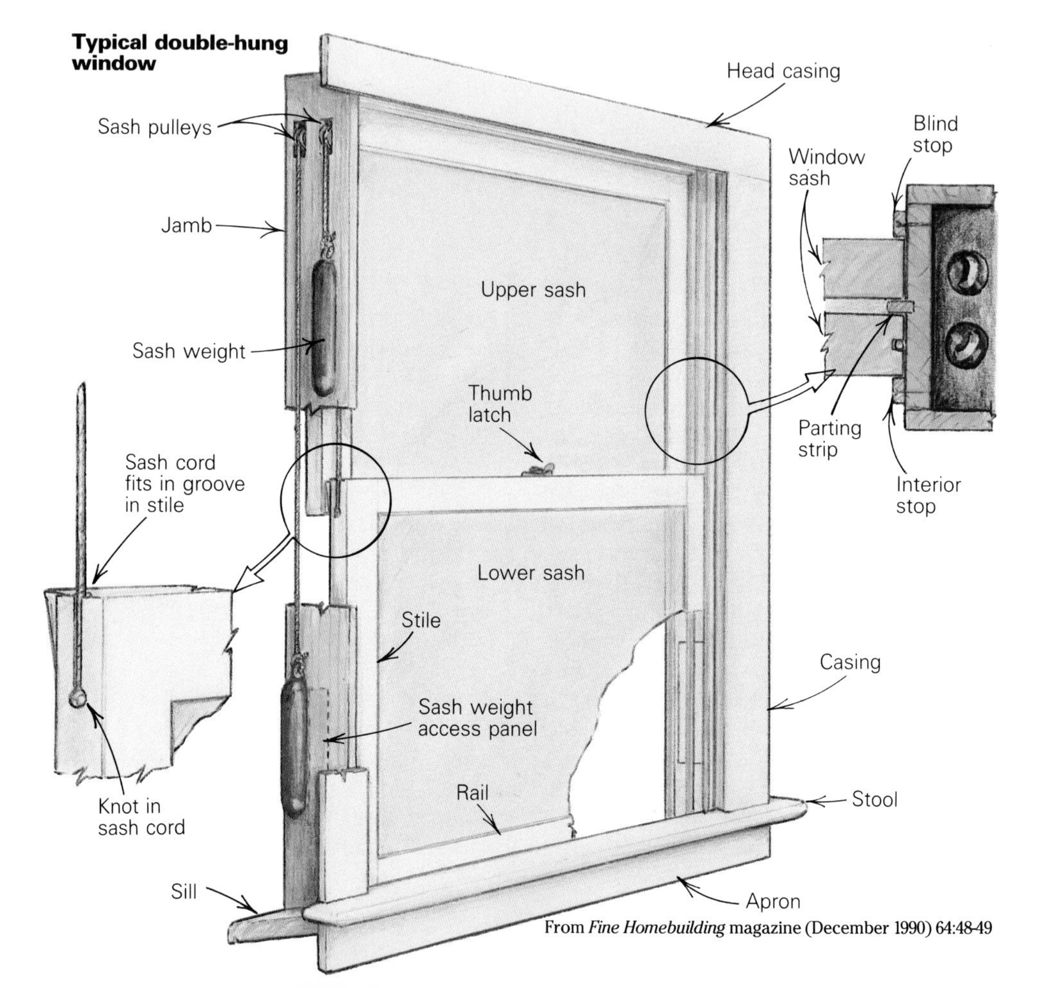

From *Fine Homebuilding* magazine (December 1990) 64:48-49

finish nails. I located the nails and used pliers to gently rock the strip sideways at each nail (I always place thick cloth or cardboard between tool and wood to prevent marring). Then I slowly pulled it straight out.

The lower sash was finally ready to be removed. Holding it by the rails, I angled its left side into the room and pulled it out of its channel on the right.

Most old double-hung windows have cotton sash cords that deteriorate over the years. Eventually they break and the sash weights drop as far as possible into the wall. Consider it good fortune if you find an intact sash cord, and treat it gingerly. I learned this the hard way—by losing the cord through the pulley. With a helper holding the window, I disengage the cord from the sash and tie the loose end around a short piece of dowel. I repeat for the upper sash. The next step with these windows was the removal of both screen sashes.

Finding the lost weights—Many window frames have access panels, making it a simple process to remove the cover plate and retrieve a weight that has parted company with its sash cord. The weight should be resting on the top of the interior sill extension.

Other window frames, however, have no access panels. Sometimes you can pull the casings off the window frame to get at a lost weight. But that wasn't the case here. The casings were narrow strips that had been plastered in place. As a result, I had to cut access holes in the window jambs (top photo).

Before cutting the holes in the jambs, it helps to calculate the length of the sash weight. Knowing the length tells you how far the access hole needs to extend above the sill to reach the top of the weight. Most weights vary from 4 lb. to 10 lb., and their lengths range from 8 in. to 16 in. To find the length of a sash weight, follow this formula: $\frac{4}{5}$ x sq. ft. of sash = pounds of each weight; $1\frac{1}{2}$ x pounds of weight = length of weight in inches (based on weights $1\frac{3}{4}$ in. in diameter). The poundage is usually stamped in Roman numerals on the side of the weight.

As shown in the photo, I positioned the 2-in. by 14-in. access hole so that most of it is concealed by the inner stop. After drilling pilot holes at the corners of the layout, I used my jigsaw to make the cuts in the jamb.

Once the hole is opened up, the weight may be visible. Use a heavy coat hanger to snag it by the eye. If you don't see it, use the wire as a probe. It is often possible to hear metal on metal and to hook the weight. If these maneuvers are not fruitful, use a small mirror and narrow-beam flashlight tied to a string to search the dark recesses.

Restringing the weights—My pulleys were brass-plated steel, and after 65 years they were still in excellent condition. A little cleaning and a couple of drops of machine oil on each shaft got rid of the squeaks and returned the pulleys to smooth working order. This kind of pulley is held fast by a couple of screws, and can be removed from the sash-side of the window. Cheaper pulleys are often press-fit into the jamb and secured from behind with a pin. They are typically serviced by removing the side casings.

To retrieve the sash weights, the author cut access holes in the window frames. When reassembled, the hole will be plugged and almost completely obscured by the interior stop. Here Strawderman reinserts a sash weight into its chase after attaching a new cord to it.

Changing cords. **We pulled this window's casing to reveal the delicate operation. The sash weight is secured with a screwdriver as the old knot is unraveled using needle-nosed pliers.**

I replaced the old cords with the best nylon-reinforced cotton sash cord that I could find. That is Magnolia Sash Cord (Wellington Leisure Products, Inc., 1140 Monticello Rd., Madison, Ga. 30650; 404-342-1916). When replacing sash cord, it's important to avoid a cord with a separate center core—the outer sheath will often wear through prematurely.

The cords to the upper sash are installed first. Before replacing each pulley, I tied a fishing sinker to one end of the sash cord and lowered it through the pulley mortise to the newly cut access hole. Sinker removed, the sash weight is tied to the bottom end of the cord, while the top end needs to be threaded from the backside through the pulley. A pair of needle-nosed pliers are good for grasping the cord. Before replacing the pulley in its mortise, I tie a half-hitch knot in the end of the cord to keep it from running through the pulley by accident. Now the weight can be reinserted into its channel.

Weights still attached to original cords should be lifted until the eye of the weight is visible through the pulley opening. I pin the weight there with a long, thin screwdriver driven through its eye (photo below). I cut the old cord and replace it with enough cord to reach the sill, allowing an extra foot for attaching it to the sash.

Reinstalling the sashes—Chances are good that the sashes need repainting, and it's best to prep them before putting them back in their frames. This is also an excellent opportunity to clean and sand the frames. I apply a clear wood sealer (The McCloskey Corporation, 7600 State Rd., Philadelphia, Pa. 19136; 800-345-4530) to the unpainted portions of the frames, stops and the edges of the sash.

Make sure you have a helper for the next step. Place the upper sash in its lowered position. Now pull the cord opposite the removed stop until the weight's eye touches the back of the pulley; then let it down about an inch. Determine where the cord fits into the groove in the sash stile, tie a double knot and cut away the excess cord. Have one person hold the cord between the pulley and the sash. Now insert the cord into the slot, place the sash between the blind stop and the parting strip with the sash's edge touching the frame. Carefully release the cord. Pull the other side of the sash into the room and secure the second cord. Raise the sash to the closed position, and repeat the process for the lower sash.

Once the sashes were all back in their frames, I replaced the weight-access panels. I used small wooden wedges to hold them in place, and caulked the saw kerf. After installing the interior stops, I ran a piece of paraffin wax in the channels made by the stops and the parting strips to help the windows run silently and smoothly. □

David Strawderman is a carpenter working in Los Angeles, California.

Sources of supply

If you find yourself in need of replacement parts for old windows, you'll find a number of companies that make or distribute parts. The best guide to who has what is *The Old-House Journal Catalog* ($15.95 from The Old-House Journal Corp., 435 Ninth St., Brooklyn, N. Y. 11215; 718-788-1700). It's a good resource to have on hand for other old-house goodies, too. One of the most comprehensive sources of parts is Blaine Window Hardware, Inc. (1919 Blaine Dr., Hagerstown, Md. 21740; 301-797-6500). Their catalog ($2 to homeowners, free to home-building professionals) is filled with measured line drawings that will help you close in on just what you need.

Drawing: Bob Goodfellow

Designing and Building Leak-Free Sloped Glazing

Error-free projects demand careful detailing

by Fred Unger

Water has an uncanny ability to work its way into the most unexpected places, while the sun can quickly drive all life from an improperly designed sunroom. That's why sloped-glazing units—such as those in sunrooms, conservatories and large skylights—are among the most complex parts of the building envelope. They're also among the hardest to build correctly.

But with the right materials, careful attention to detail and a sense of how water behaves, well-functioning skylights and sunrooms are well within reach of most knowledgeable builders. Building a high-quality unit isn't cheap, but it will cost less over the long haul than endless callbacks or having to replace an entire structure later on. I tell sunroom clients that, while I can build a conventional small addition for $100 to $120 per square foot, they should plan to spend at least $160 per square foot for a comfortable, well-built sunroom with overhead glass.

Years of trial, error and refinement have passed since I worked on my first sloped-glazing project in 1974. I've learned that the secret to controlling water problems is a dry glazing system that depends not on caulks or sealants, but on gravity and physics. Before going into the details of the system, I'll review some general principles that apply to all wood-frame sloped glazing.

Custom or manufactured?—We custom-build many of our sloped-glazing units on site or in our shop. But why custom-build when there are plenty of manufactured products on the market? Part of the answer is that some glazing systems available to custom builders are superior to those found in most manufactured units (I can't understand why manufacturers have been so slow to use them). Although we do install some manufactured systems, we only work with companies that will customize their units to our specifications and with units on which we can install our preferred glazing system. To save time in the field and enhance quality control, we also preglaze some skylights in our shop, truck them to the site and hoist them into place with a crane. Most manufactured systems are designed to be site-assembled; thus, they aren't engineered to withstand the added stresses of transport and hoisting.

The other part of the answer concerns design. Architects and clients can be quite creative in their designs (for example, we recently completed an African mahogany pyramid skylight with heat-mirror glass), but there are limits as to how far manufacturers will go to customize their products. So far, we haven't found a manufacturer that meets all of our demands for design and detailing.

On architect-designed projects, we try to get involved early in the design process. By sharing our expertise, we can help specify the framing materials, glass and sealants. We also suggest details that will expedite the construction process and save the client money. Our goal is a durable, efficient, cost-effective structure that meets the architect's and the client's design goals. Offering such help has gotten us specified into projects without competition.

Choosing a frame—The most critical design factor we must deal with is movement. On a sunny winter day, the components of a south- or west-facing skylight or sunroom may be exposed to some extreme temperature changes. The differing rates of expansion and contraction of wood, metal, glass and rubber can lead to buckled flashings, torn caulk joints, failed water seals and even broken glass. The movement is even more pronounced in poorly designed, wood-framed spaces that house plants, a pool or a hot tub. I've seen sunrooms built with construction-grade lumber in which the frames had twisted or warped well over an inch.

Because movement is so potentially damaging, a dimensionally stable frame is a must (photo right). The use of unstable wood can lead to major problems even if all other aspects of the project are executed perfectly. We prefer to work with clear, kiln-dried, vertical-grain stock. The most dimensionally stable

Materials count. **A skylight or sunroom must withstand wind, rain and sun, so even a simple project demands a dimensionally stable frame. The best domestic woods are clear, vertical-grain redwood and cedar (photo below). Outside, hips, valleys and angles can be covered with custom aluminum caps that can be fabricated at a sheet-metal shop (photo right). They're held in place by exposed stainless-steel screws with gasketed washers.**

species we've found are redwood and cedar, although we've also had good luck with some species of mahogany. Good wood can be quite expensive. Luckily, however, laminated wood is also an excellent choice: It's less expensive and more environmentally friendly than solid stock cut from old-growth trees. It's also more stable. We never use solid sawn hem-fir, white pine, yellow pine or oak except as part of a composite member.

Regardless of the species, the best glazing systems we've found require the surface on which two pieces of insulating glass come together to be about 3 in. wide. The bearing area can consist of solid wood or of glued 2x or 1x stock. If these pieces will be exposed to high humidity, a two-part resorcinol or other waterproof glue should be used.

Precision and flexibility—We try to design our frames to meet two seemingly contradictory demands: precision and flexibility. The joinery on most of our projects is complicated, especially when hips and valleys are included. Because virtually every cut ends up as exposed finish work, our frames have to be precise. Structural connections demand careful planning and intelligent detailing. Once ordered, expensive insulating glass can't be trimmed to fit improperly framed openings. Because of this, we must design and build our frames to close tolerances.

We usually join our framing components with blind screws, dowels, biscuits or other hidden fasteners. We've also used exposed brass screws. Some areas, such as structural ledger boards, can be hidden. We attach them with nails or screws and then hide them with trim. Our choice of fasteners depends on the expected humidity level and the wood species we're using. In a high-humidity environment, the natural extractives in cedar and redwood will corrode even galvanized fasteners. We tend to use a lot of stainless-steel screws and nails.

Despite the fact that our frames must double as finish work, they also have to perform in the real world of construction. We can't expect framers, masons and foundation crews to work to 1/16-in. tolerances in three dimensions. We compensate by designing shim spaces into the system and then use trim to span the gaps.

The system—Many glazing systems rely on caulks and sealants to prevent water problems. But we've found sealant-dependent systems to have some real disadvantages. While some of them will stop leaks if installed under near-ideal conditions, most won't control condensation. These systems also make glass replacement difficult. I've spent hours with scrapers and solvents, removing broken glass that had been caulked in place. But it wasn't until a client's dog stepped on a gob of urethane caulk and tracked it across a $5,000 rug that I became determined to minimize our use of these products.

The dry glazing systems we now use employ rafter and purlin baseplates with internal gutters, EPDM rubber gaskets, insulating glass and a gasketed aluminum pressure cap (drawings right). The aluminum cap holds the glass in place. EPDM gasketing or closed-cell foam tape serves as the primary water seal. A removable trim cap hides the screws, which are installed with gasketed washers. On purlins, we use flat bar stock over closed-cell glazing tape because the resulting low profile prevents water damming. The glass sits on raised pads that prevent any water in the guttering system from puddling against the window's edge sealant. Such puddling is a leading cause of seal failure and fogging within insulated glass. To keep the glass from sliding and to keep adjacent pieces of glass from shearing relative to each other, we rest its bottom edge on rubber setting blocks and aluminum setting-block supports. These blocks support the glass and keep top and bottom lites from shearing relative to each other. The supports are positioned one quarter of the way in from the bottom corners of the glass.

But the guts of this system are the internal gutters in the baseplates. Good systems are designed to direct water from the purlin gutters to the rafter gutters to weep holes at the eaves. In effect, the gutters serve as a safeguard to the primary water seal. As one old-time glazer told me, "If you don't want water problems, you've got to design the system to leak." The dry system also permits quick, low-cost replacement of broken glass.

We use these systems because we consider them to be the most foolproof. Even though we'd like to pretend that our designs, our installers and our materials are perfect, they're not. For best results, workers should understand how water moves, along with why and how the system sheds water. Fortunately, however, the details we've developed will work even if installed on a Monday morning. On one job, we set the glass and then got a solid week of rain before we could cap the system off. We didn't get a single leak.

Baseplates can be aluminum with EPDM gaskets, or solid EPDM rubber (photo page 72). The best we've found are available from Abundant Energy, Inc. (P. O. Box 307, County Rt. 1, Pine Island, N. Y. 10969; 800-426-4859) and U. S. Sky (2907 Agua Frio, Santa Fe, N. M., 87501, 800-323-5017). The aluminum systems have been around for several years and many now include thermal breaks. Aluminum baseplates hold screws for the pressure caps more securely than wood systems do, and ensure a flat surface for glazing. But they take more time to install than the rubber ones and are hard to work around hips and valleys without adding cumbersome-looking details. If they don't include a built-in thermal break, aluminum baseplates themselves can cause minor condensation and interior frosting. Even thermally broken aluminum systems must be carefully detailed to prevent thermal conduction problems.

The EPDM baseplate, on the other hand, installs quickly and easily (it cuts with shears or

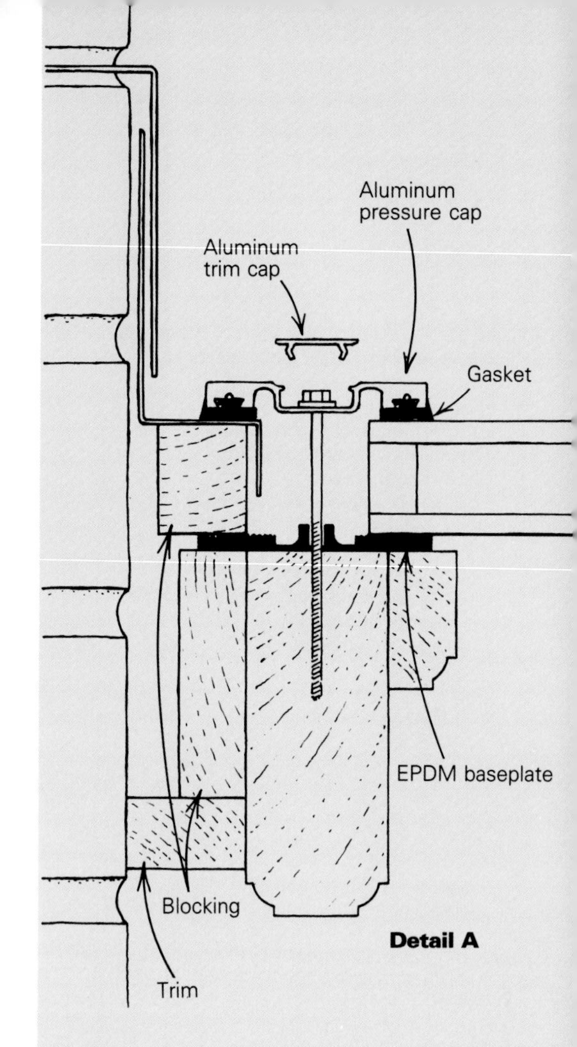

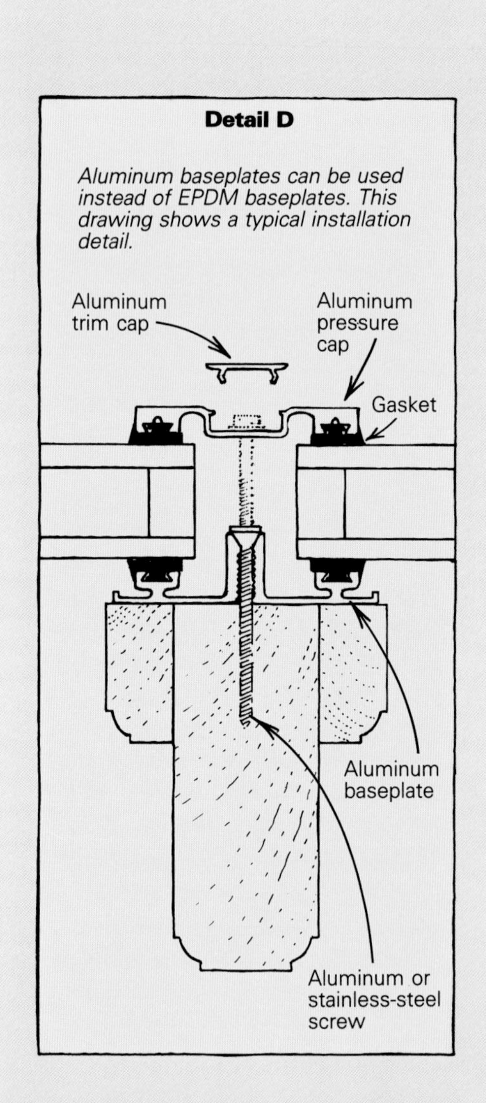

Aluminum baseplates can be used instead of EPDM baseplates. This drawing shows a typical installation detail.

From *Fine Homebuilding* magazine (February 1992) 72:76-80

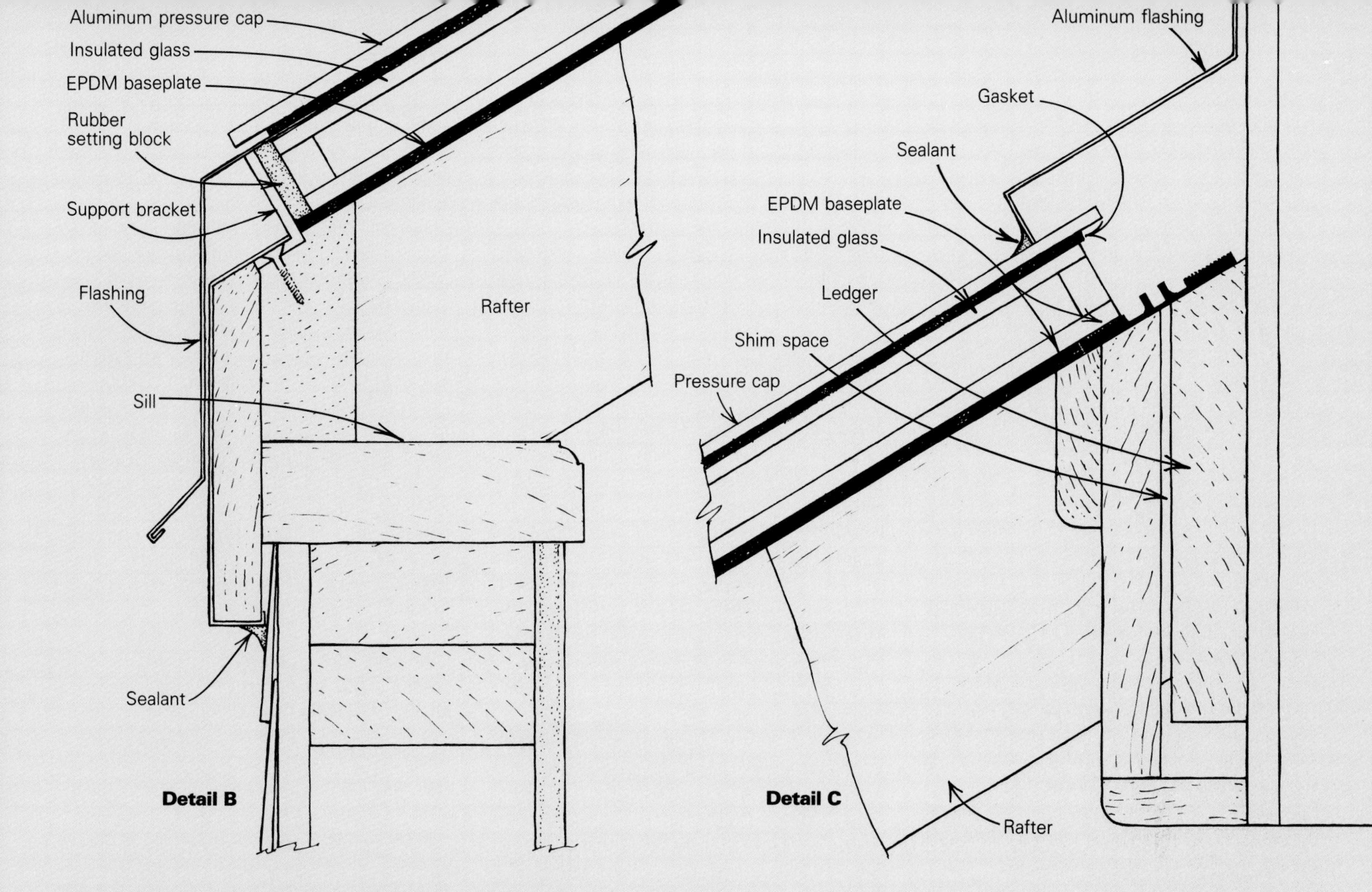

Shedding water. *A dry glazing system is designed around a system of aluminum or EPDM rubber baseplates with internal gutters. By directing any leaks or internal condensation to weep holes at the eaves, the gutters serve as a back up to the primary water seal. The system also includes EPDM rubber gaskets, insulating glass and an extruded-aluminum pressure cap. The aluminum pressure cap holds the glass in place. Raised pads beneath the glass prevent any water in the gutters from puddling against the window's edge sealant. The bottom edge of the glass rests against setting blocks and setting block supports. In addition to shedding water very effectively, the system also permits quick, low-cost glass replacement.*

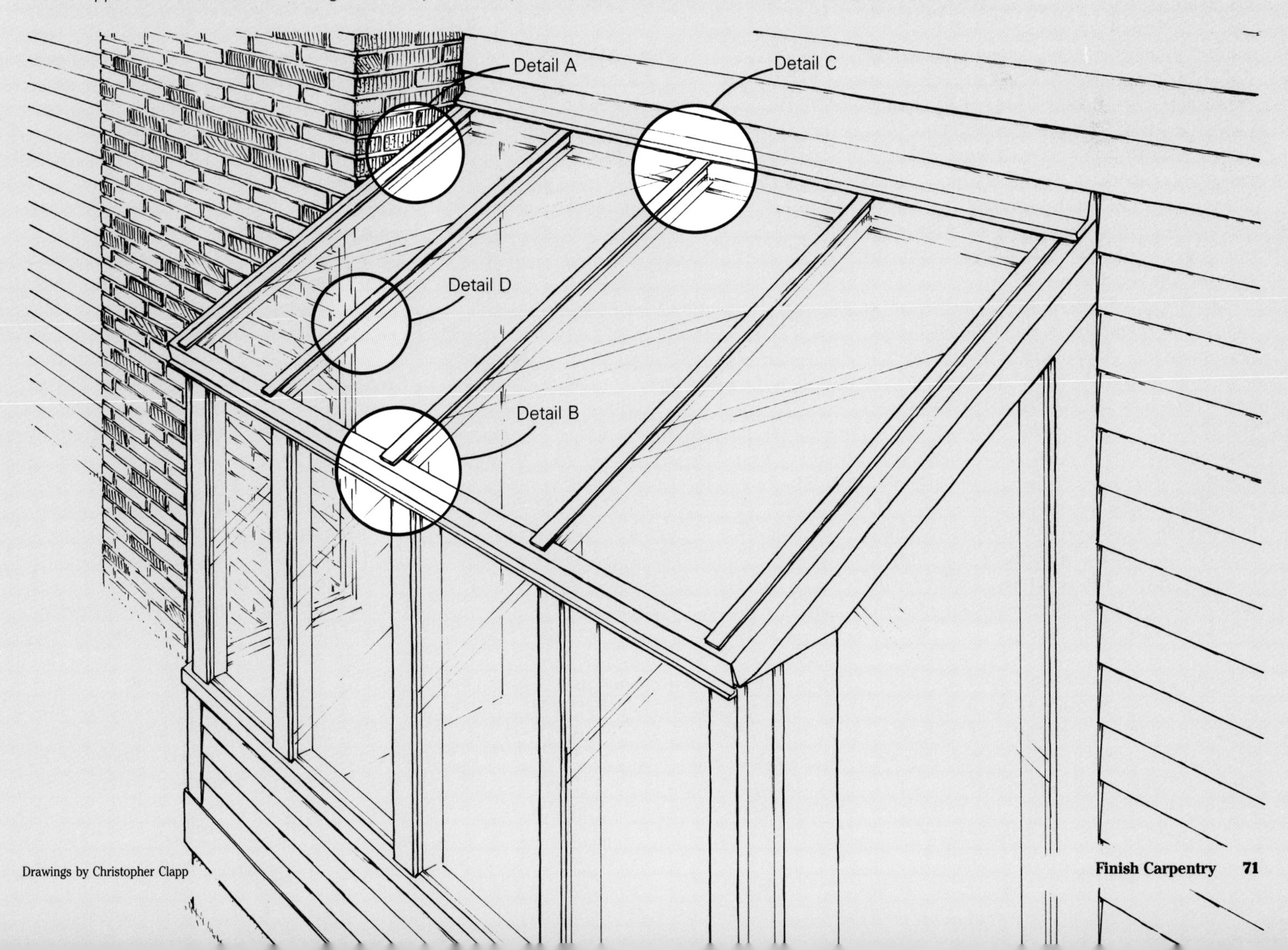

Drawings by Christopher Clapp

a utility knife), bends around hips and valleys and isn't prone to condensation or frost problems. Unfortunately, the EPDM baseplates that have been available don't have condensation gutters and cannot be easily installed to direct water from a horizontal purlin to a vertical rafter. They should only be installed on systems with one lite of glass vertically. Even then, the glass has to be carefully detailed.

Abundant Energy is about to introduce an EPDM baseplate that incorporates cascading internal gutters. It's fast and flexible to install and can be cut with a utility knife, yet it costs less than the aluminum system. It also has the fastest-to-install setting block support of any system I've found on the market, isolates water from any penetrations through the baseplate and eliminates any concern about thermal conduction.

Metals and flashing—Even with rubber baseplates, metals remain a crucial part of any sloped-glazing project. Baseplate gutters aren't intended to handle rivers of water, so it's important to have good exterior cap and flashing details. We clad the exterior sloped portion of all our projects with an aluminum cap (photo, page 69). Abundant Energy and U. S. Sky have excellent caps, as do most manufacturers of sunroom kits and curtain-wall glazing systems. Most of these caps are designed for use where two lites of glass meet on a flat plane. For hips, valleys and angles, we have a sheet-metal shop fabricate custom caps from heavy aluminum flat stock with an anodized or baked-enamel finish. These must be fastened using exposed stainless-steel screws with gasketed washers.

Occasionally, an architect will insist on copper or lead-coated copper flashings. But copper and aluminum are galvanically incompatible; using them together raises the possibility of corrosion (for an explanation of galvanic corrosion see *FHB* #62, pp. 64-67). When we have to put galvanically incompatible metals in close proximity, we make sure to isolate them with wide, closed-cell foam tape or with EPDM gaskets.

Metals expand and contract greatly with changes in temperature. Before screwing or nailing through caps or flashings, we always predrill or punch over-sized pilot holes. To protect against galvanic reactions, we use stainless-steel fasteners or fasteners of the same metal as the flashings. Prepainted aluminum flashings also can't be soldered. Where two or more lengths of flashing are required for a single run, we leave a ⅛-in. gap between them and then install a spline below the gap. When the edge of the flashing is bent over and splined, it locks the two runs together.

Sealants—Despite our best efforts to minimize them, we still use lots of sealants (for more on caulks and sealants see *FHB* #61, pp. 36-42). To prevent leaks, we always run a bead of silicone along the top edge of our purlin caps and eaves flashings, then extend it a couple of inches up the edge of the adjacent rafter caps. We also use sealants at butt joints in flashings. We use sealants that will permit high levels of joint movement; this is especially important when transporting preglazed skylights over bumpy roads.

At one time, we tried using only neutral-cure silicones. They're easily gunned, permitting a nice, clean bead, and they were the only silicones I could find that showed any tenacity in sticking to wood. Most silicones also adhere well in glass-to-metal connections. But we found that regardless of how well they adhered when first applied, the silicones would eventually begin to pull away from wood. We now use urethane sealants on all wood and masonry joints. The urethanes are absolutely remarkable in their adherence, though they're not as easily gunned or tooled as the silicones. We still use silicones for glass-to-metal connections (such as the leading edge of the eave flashing) or for metal-to-metal connections (such as the intersection between two flashings).

Pros and cons. **Baseplates can be aluminum or EPDM rubber. Aluminum provides a stable bearing surface for glazing but can be time-consuming to install, and it is subject to interior frosting. The EPDM doesn't frost up but doesn't have a condensation gutter. Photo by Susan Kahn.**

One thing to watch for when choosing sealants is potential chemical incompatibility between the field-applied products and the insulating glass-edge sealants. The chemical reactions between two incompatible sealants can lead to the breakdown and failure of one or the other. Several builders, glass companies and window manufacturers have been ruined by lawsuits that resulted from chemical-related seal failures. Your insulating-glass manufacturer should be able to provide sealant-compatibility test results upon request.

Caulks and sealants should only be applied to clean, dry surfaces. Before caulking, we clean adjacent surfaces with glass cleaner. To cut any oils or residual films, we usually wipe the joint with a rag dampened slightly with a solvent called xylol. This is also the best solvent we've found for removing silicone—which is why we take care to keep it away from the window edge sealants.

Choosing the glass—Insulating glass consists of two or more lites of glass separated by an air- or gas-filled space. A hollow aluminum spacer is joined to both laminations with a moisture-proof edge sealant. The aluminum spacer is filled with a dessicant that keeps the air between lites dry.

When ordering glass, be sure to check its warranty, code-compliance and thermal and solar characteristics. I'll touch on the first two here; the third is a subject for a future article. The glass used in a sloped-glazing unit should be guaranteed for use on a slope (most glass isn't). Putting glass on a slope adds differential stresses between the inboard and outboard lites that aren't present in vertical glazing. Most sloped-glazing units have a dual-edge sealant—a silicone structural sealant and a moisture seal consisting of some other material. A few companies are switching to a new single-seal urethane. If you're using urethane sealed units, it's important to ensure that all edges of the glass are capped. The sealant will break down if it is exposed to direct sunlight.

The second major consideration with sloped glass concerns safety and codes. Most codes require that sloped glass on commercial buildings include an outboard layer of tempered glass and an inboard layer of laminated safety glass. The tempered glass is strong enough for someone to stand on and usually won't break unless tapped on an exposed edge or hit with a sharp object. The plastic layer within laminated safety glass is strong enough to hold any tempered glass that does break, along with whatever broke it. These safety standards make sense for residential work, too; in fact, several states have incorporated the standards into their residential building codes. □

Fred Unger is the owner of Heartwood Building Specialties in Berkeley, Mass. He designs and builds custom additions, skylights and sunspaces. Photos by Charles Wardell, except where noted.

Wooden Miter Boxes

A traditional approach to cutting corners

by Tom Law

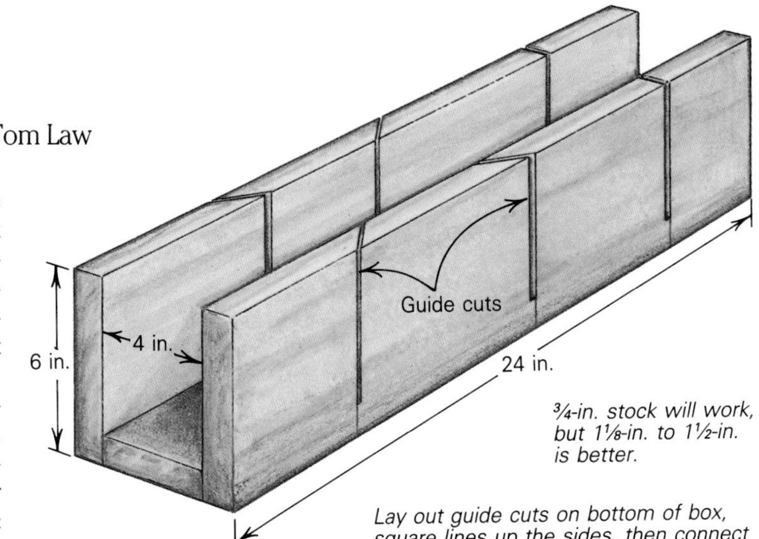

The electric miter saw is the best power tool to come along since the circular saw, and like all power tools, it is best at doing repeated work with speed and accuracy. It is not the only way to make miters, however. Wooden miter boxes were used long before electric ones were invented and can do a lot of work before wearing out. Moving from framing to trim work requires that you slow the pace, so a good way to start is by making a miter box.

The wood you choose should be close-grained and free of knots because when the saw guides are cut, the sides must remain straight. White pine and oak are the usual choices but birch, maple and mahogany are also used. The size of the box, as well as the angle and number of guide cuts, will vary according to the kind of trim to be cut. A basic size would be 24 in. to 32 in. long, 4 in. wide on the inside and 6 in. high (drawing right). The sides can be ¾ in. thick, but heavier stock (1⅛ in. to 1½ in.) will work better. Make the sides higher than the trim so the saw can enter the guides above the trim. The edges of the bottom piece must be perfectly square; when the sides are attached they should form a right angle with the bottom. I prefer to glue the sides in place, but they can also be fixed with screws or nails.

When the sides have been attached, turn the box upside down to lay out the guide cuts. Place a framing square across the bottom and draw a 45° line across the width of the box. Turn the box on each side and square the marks up, then turn the box upright and lay the framing square across these marks to double-check for accuracy. Mark the lines across the top edges, then square them down the insides.

Begin cutting the guides by placing the handsaw on the top edge of one side and to one side of the line. Don't cut the pencil lines out—they should be visible when the cut is finished. Drag the saw toward you to start the cut, then slowly use longer and longer strokes. Cut only a short way down, then angle the point of the saw up and cut a little farther. When the cut is about halfway down on one side, shift the box around and repeat this procedure on the other side. Now put the saw through both sides and saw both at the same time. When you are sure the saw guides are in line, cut again on only one side but allow the saw blade to remain at the opposite guide. Again shift the box around and saw that side, then finish up by sawing both at once. Each set of guides should be made this way.

Miter boxes can also be made with just one side (photo right). Small moldings can be accurately cut with this kind of box. Another type of one-sided box, which I use to cut wide base molding, is simply a saw guide nailed to the cross beam of a sawhorse. The advantage of a one-sided box for baseboard is that the handling of the material is easier.

It is important that the saw you use in the miter box be properly sharpened. If all the teeth are of equal height, the saw will go right down the line. If the teeth on one side of the saw are high, they will cut first and the saw will wander from the line.

Although backsaws are traditionally associated with miter boxes, I prefer to use a standard carpenter's handsaw. Backsaws are too short or too heavy or back home when I need them on the job site. A properly sharpened handsaw will work just as well, or even better. The finer the teeth, the finer (and slower) the cut will be. Set also affects smoothness; a wide set makes a rougher cut. I like to use a 10-point saw with very little set because it makes as smooth a cut as an 11-point saw, but makes it faster. The saw used to cut the guides should be the only saw used in the miter box. The saw is guided by the cut it has made and should fit snugly. Widening the guides by using a coarser saw will allow a finer saw to chatter from side to side, gradually widening the guides enough to affect the accuracy of the box. A little paraffin rubbed on the sides of the saw will help it to glide along and won't stain the trim being cut.

Law finds a one-sided miter box adequate for cutting small moldings. The sawhorse with a 2x8 nailed to the top is a bigger version of the same thing, but is used for large baseboard.

When using the miter box, your motion should be free and easy. Put all your weight on your feet and position your body so that your arm travels in line with the cut. The stock should be against the back wall of the box to resist the pushing of the saw and should be held firmly in place by your free hand. Grasp the saw handle with just enough pressure to hold on to it. Push and return the saw with no downward or sideways pressure; the weight of the saw will be enough to carry it through the cut. Sometimes a piece has to be shortened by just a saw cut. To do that, take a scrap piece and tack it to the bottom of the box right at the kerf line. Resistance added by the block will allow the saw to cut without wandering to the free side. □

Tom Law is a carpenter in Westminster, Maryland, and a consulting editor of Fine Homebuilding. *Photo by Kevin Ireton.*

Drawing: Bob Goodfellow

From *Fine Homebuilding* magazine (February 1991) 65:75

Plate Joinery on the Job Site

Quick and easy insurance against joints opening up

by Kevin Ireton

Like most people when they first buy a plate joiner, David Mader, a carpenter in Yellow Springs, Ohio, wanted to find out how strong plate joints really are. Mader crosscut a 2x4 and reassembled it with a pair of no. 20 plates (the largest size available), one over the other. After letting the glue dry, Mader tried to break the 2x4 over his knee. He couldn't do it. Convinced of plate joinery's strength, Mader proceeded to use his plate joiner to butt-join custom flooring that wasn't end-matched.

Often considered the province of shop-bound woodworkers, plate joinery, it turns out, is being used more and more by carpenters on the job site (photos at right). Plate joinery and biscuit joinery are the same thing, and in this article I'll use the terms interchangeably.

The basic idea behind plate joinery is simple: plunge a 4-in. circular sawblade into a piece of wood, and you get a crescent-shaped slot. Make a series of these slots along the edges of two boards that you want to join. Insert glue and a football-shaped wooden spline into each slot on one board. Insert more glue into each slot on the other board, then press the two boards together. Water from the glue causes the splines to swell, making a strong, tight joint.

Biscuit-joiner basics—The typical biscuit joiner is a cylindrical machine (drawing facing page), about 10 in. long, and weighs between 6 lb. and 7 lb. It has a D-shaped handle on top and a spring-loaded faceplate in front with an adjustable fence. When the tool is pressed against the workpiece, a 4-in. carbide-tipped blade extends through a slot in the faceplate and scoops out a kerf in the workpiece. You can adjust the distance between the kerf and the face of the workpiece, but any closer than 3⁄16 in., and the biscuit, or plate, may pucker the surface of the wood when it swells. You can also adjust the depth of the kerf to fit the size of biscuit you're using.

Biscuits come in three basic sizes (all three are arcs of same circle): #0 is about 5⁄8 in. wide and 1¾ in. long, #10 is ¾ in. wide and 2⅛ in. long, and #20 is 1 in. wide and 2½ in. long. Biscuits are made of beech with the grain oriented diagonally to the length, making them very strong across their width. Biscuits are also compressed so they'll fit easily in the kerf and then swell once the glue hits them. All biscuits are slightly shorter than the

Slotting in place. **To deal safely with a small piece, this carpenter installed the plinth block first and slotted it in place (above). In the photos below, he adds a biscuit, then the casing.**

kerf they fit into, which not only allows room for excess glue but also provides some play for aligning a joint along its length. This gives biscuit joinery a distinct advantage over doweling as an indexing technique.

Plate joinery works in hardwood, softwood, plywood, particleboard and even in solid-surface countertops (using Lamello's clear plastic C-20 biscuits). Plates can be used in edge-to-edge joints, butt joints and miter joints.

Joinery comes to the job site—Over the past 15 years, plate joinery has proven itself strong enough and accurate enough to earn a place in many woodworking shops, where it competes with other joinery methods such as doweling, splining or mortise-and-tenon joinery. The merits of plate joinery relative to these other methods can and have been debated. But most carpenters in the field don't enjoy the luxury of a fully equipped shop, and often their only joinery options are whether to use nails or screws. That's why plate joinery adds a valuable technique to a carpenter's repertoire. After all, a cabinetmaker can successfully argue that a biscuit-joined face frame is not as strong as one joined with mortises and tenons. But no one will argue that adding a biscuit between two pieces of mitered casing (photos, p. 76) won't strengthen the joint or greatly improve its chances of weathering changes in humidity without opening up.

Joint strength isn't the whole story, though. A biscuit joiner is very portable, taking up less room in a toolbox than a circular saw. And it's extremely fast. Cutting slots and adding biscuits to a mitered door casing might require 30 seconds. Admittedly, even that little time can be significant when multiplied by a houseful of doors and windows. You might consider it worthwhile, though, if you've ever been disappointed when returning to a job to discover gaps in joints that fit perfectly when you nailed them up.

Who's using them where?—Stephen Sewall, a builder in Portland, Maine, feels so strongly about the advantages of biscuit-joined trim that he seldom installs trim without biscuits. On a recent job where he didn't have his biscuit joiner, Sewall nailed up the side casings, but left the head casings loose so that he could add the biscuits later.

Sewall also says that biscuit joinery has

Drawings: Bob Goodfellow

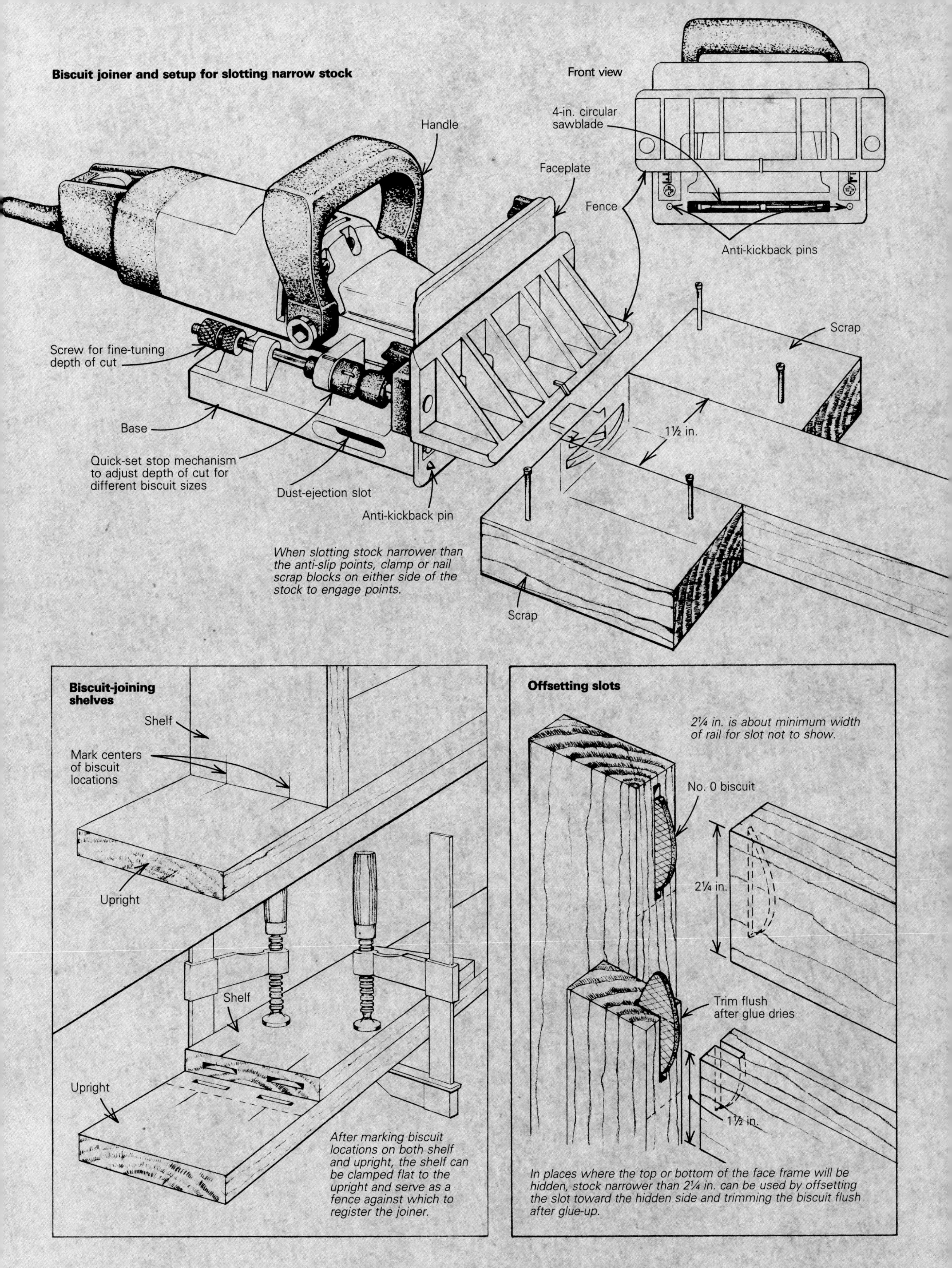
Biscuit joiner and setup for slotting narrow stock
Front view
4-in. circular sawblade
Handle
Faceplate
Fence
Anti-kickback pins
Scrap
Screw for fine-tuning depth of cut
Base
1½ in.
Quick-set stop mechanism to adjust depth of cut for different biscuit sizes
Dust-ejection slot
Anti-kickback pin
When slotting stock narrower than the anti-slip points, clamp or nail scrap blocks on either side of the stock to engage points.
Scrap
Biscuit-joining shelves
Shelf
Mark centers of biscuit locations
Upright
Shelf
Upright
After marking biscuit locations on both shelf and upright, the shelf can be clamped flat to the upright and serve as a fence against which to register the joiner.
Offsetting slots
2¼ in. is about minimum width of rail for slot not to show.
No. 0 biscuit
2¼ in.
Trim flush after glue dries
1½ in.
In places where the top or bottom of the face frame will be hidden, stock narrower than 2¼ in. can be used by offsetting the slot toward the hidden side and trimming the biscuit flush after glue-up.

It only takes a second. **With the tool and the trim registering against the floor, this carpenter makes short work of cutting slots. By then adding a biscuit spline between two pieces of mitered casing, he prevents the joint from opening as a result of wood shrinkage.**

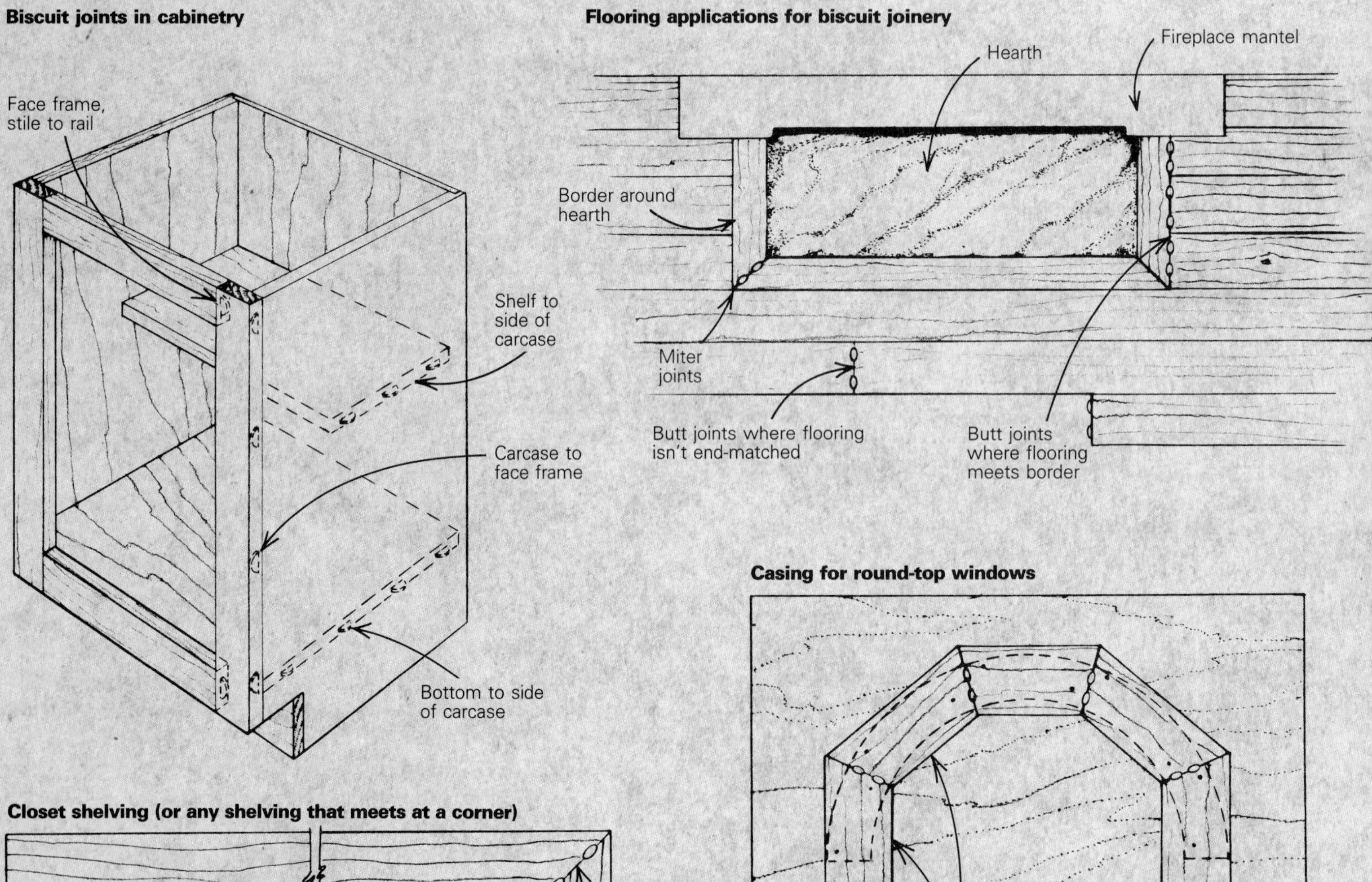

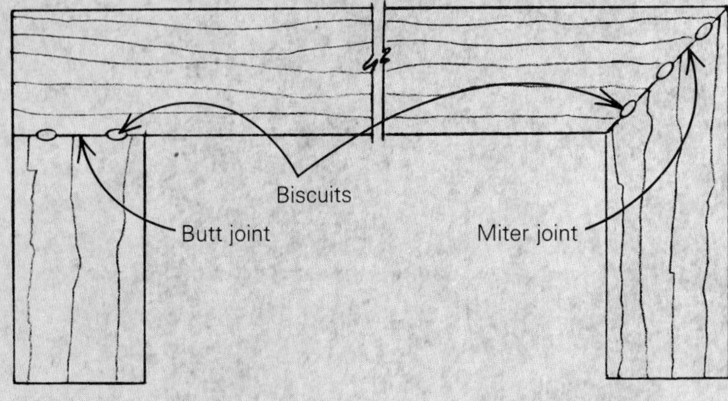

made building cabinets on site a lot easier (drawing facing page). When installing a fixed shelf in a cabinet, which he used to do by routing dados in the sides to house the ends of the shelf, Sewall can now biscuit-join the shelf to the carcase faster than he can change bits in his router. When biscuit joining shelves, Sewall often clamps the shelf flat against the upright and registers the biscuit joiner against it as he cuts the slots (see drawing, p. 75). Biscuits will also work in ½-in. stock, like the ½-in. Baltic-birch plywood that Sewall uses to make cabinet drawers.

Foster Jones, a partner in Maine Coast Builders of York, Maine, admits that using biscuit joinery adds to the cost of a job and says his company usually decides before starting a project whether to use biscuits. If they do use them, though, they don't just use them on miter joints. They use biscuits to join inside and outside corners of baseboard, to join baseboard to door casing and to join door casing to plinth blocks (photos, p. 74).

When laying a hardwood floor, the carpenters at Maine Coast Builders use biscuits to join the picture-frame border around a fireplace hearth (drawing facing page) and to join the border to the flooring that butts into it. They even use the biscuit joiner as a trim saw to trim the bottoms of door casing so that flooring will fit beneath it.

Jones also uses biscuits when fabricating trim for round-top windows. Using biscuits and five-minute epoxy, he joins mitered segments of straight stock end to end in a rough semicircle (drawing facing page). He usually screws the segments to a piece of plywood rather than clamping them. The five-minute epoxy lets Jones work with the piece after less than an hour of drying time.

Because biscuit joinery relies in part on the biscuits' capacity to absorb water from the glue and swell up to form a tight joint, you may be wondering how well the system works with epoxy, which isn't a water-based glue. Jones and Sewall wondered, too, and broke apart joints that they had assembled with epoxy. Both found the joint to be just as strong as those made with yellow glue, which has a water base. In fact, Bob Jardinico at Colonial Saw, sole U. S. distributor for Lamello joiners (see the sidebar at right for address), recommends epoxy for biscuit joinery used outdoors. Makes you wonder if biscuits and epoxy aren't the way to keep mitered handrails on exterior decks from opening up.

Plate-joining face frames—A common complaint when assembling face frames (or cabinet doors) with biscuit joints is that the rail must be wider than the slot for the smallest biscuit. Otherwise the biscuit will show. This limits you to a rail that's at least 2¼ in. wide. Responding to this complaint, Lamello recently introduced face-frame biscuits (H-9) that are ½ in. wide by 1½ in. long. But such a short biscuit means you have to switch to a 3-in. sawblade (also available from Lamello).

A company called Woodhaven (5323 West Kimberly Rd., Davenport, Iowa 52806; 319-391-2386) makes biscuits out of particleboard that are 1-in. wide by 1⁵⁄₁₆ in. long and for which you cut kerfs with a 6mm slot-cutting router bit (available from Woodhaven and from MLCS, Ltd, P. O. Box 4053, Rydal, Pa. 19046; 800-533-9298). With these same bits you can also use your router to perform conventional plate joinery, but the cutter is exposed and you don't have a faceplate, so you lose some of the safety and convenience of a plate joiner.

As it turns out, though, you can often get away with using a standard biscuit in a narrow rail by offsetting the slot to the outside of the frame and trimming the biscuit flush (drawing, p. 75). In most cases the exposed kerf and biscuit are either pointing down at the floor and can't be seen, or are covered by a countertop. When cutting slots in narrow stock, it's best to clamp or nail scrap blocks on either side so the steel points on the joiner's faceplate have something to grip (drawing, p. 75). These points keep the tool from slipping during a cut (some joiners employ rubber bumpers or pads rather than steel points).

But wait, there's more—It's easy to think of other job-site uses for biscuit joinery: mitered ceiling beams, jamb extensions on doors and windows, return nosings on stair treads. You could even use biscuit joinery where two closet shelves meet at a corner and avoid having to screw a cleat to the underside of one shelf to support the other (drawing facing page).

Do be careful, though, if you decide to buy a biscuit joiner. Beware the "Law of the Instrument." This is a theory in psychology that states: if you give a small boy a hammer, everything he encounters will need hammering. There may be some things that really don't need to be joined with biscuits. □

Kevin Ireton is managing editor of Fine Homebuilding. *Photos by the author.*

Biscuit-joiner manufacturers

Looking for a faster and more accurate alternative to doweling, a Swiss cabinetmaker named Henry Steiner developed plate joinery (also called biscuit joinery) back in the 1950s. Steiner founded a company to manufacture slot-cutting machines, called plate joiners, and the plates (or biscuits) to go with them. The company is called Steiner Lamello, Ltd. ("lamello" comes from the German word *lamelle*, meaning "thin plate"), and despite the fact that nine other companies now make plate joiners, the Lamello machine is still considered by many woodworkers to be the "Cadillac" of plate joiners. The following list includes all the companies that make plate joiners. Prices range from under $150 to over $600. So if you decide to buy a plate joiner, be prepared to shop around. —*K. I.*

Delta 32-100 (bench-top model)
Delta International Machinery Corp.
246 Alpha Dr.
Pittsburgh, Pa. 15238
(800) 438-2486

Elu Joiner/Spliner 3380
Black & Decker, Inc.
P. O. Box 798
Hunt Valley, Md. 21030
(800) 762-6672

Freud JS 100
Freud, Inc.
P. O. Box 7187
High Point, N. C. 27264
(919) 434-3171

Kaiser Mini 3-D
(dist. by W. S. Jenks & Son)
1933 Montana Ave. NE
Washington, D. C. 20002
(202) 529-6020

Lamello (3 models)
(dist. by Colonial Saw Co., Inc.)
P. O. Box A
Kingston, Mass. 02364
(617) 585-4364

Porter-Cable 555
Porter-Cable Corp.
P. O. Box 2468
Jackson, Tenn. 38302
(901) 668-8600

Ryobi JM100K
Ryobi America Corp.
5201 Pearman Dairy Rd.
Suite 1
Anderson, S. C. 29625
(800) 226-6511

Skil 1605:02
Skil Corp.
(subsidiary of
Emerson Electric Co.)
4300 West Peterson Ave.
Chicago, Il. 60646
(312) 286-7330

Virutex O-81
Rudolf Bass Inc.
45 Halladay St.
Jersey City, N. J. 07304
(201) 433-3800

Router Control

Using site-built jigs, bearing-guided bits and a router table to make precise cuts in wood

by Jud Peake

I've been using routers for over 15 years, and I'm still amazed by their versatility. Armed with a jig, a router and a sharp bit, I can hog out mortises for stair treads in a rock-hard stringer made of Douglas fir. Using the same rig mounted under the wing of my table saw, I can quickly mill a stack of delicately curved trim pieces out of expensive molding stock.

I'm a fan of big routers. Unless you plan to confine your router work to trimming the edges of plastic laminates or cutting out shallow mortises for hinges, a small router just won't cut it. You want big rpm (over 20,000), and you want big horsepower (2 hp minimum—3 hp is even better).

Not even a champion arm wrestler can freehand control a 3-hp router spinning at 22,000 rpm. You've got to restrain the tool to make sure it goes where you want it, and that's what this article is about. Someone with only a little experience with a router can get consistently perfect results using the guides and jigs that I'll describe here.

Base-guided router—There are three common techniques used to guide a router as it cuts a negative shape, such as a mortise. *Base-guiding* the router requires a corral (top drawing, facing page). On a scrap of plywood tack a perimeter fence, or corral, that will confine the router's base to a set area. The size of the corral is found by this formula: mortise width + base diameter - cutter diameter = corral width.

When I use this method I make test cuts to see if I've tacked the fences in the correct position. Once I've got them right, I screw the fences down to make sure they won't move. I think this kind of time-consuming fiddling is the drawback with the base-guided setup. You also have to make sure that the base of the router is centered on the collet, or your results will be inaccurate. Also remember that a ding in the corral will be faithfully followed in each succeeding routing.

On the other hand, the fences are well away from spinning cutters as the base-guided router is inserted and withdrawn from the work. Consider this when making any production jigs expected to have a long life.

Collar-guided router—Some routers have a collar that will screw into the router's base (bottom drawing, facing page). The collar has a flange that bears against a template as the router is moved over the workpiece. The formula for calculating the dimension of the template corral is: mortise width + collar diameter - bit diameter = corral width.

I think the collar setup has the disadvantages of the base-guided one with none of the advantages. The calculating and fiddling take just as long, yet the template is so close to the bit that it's likely to be damaged.

Bearing-guided router bits—I think the low-tech (and thus low-budget) approach appropriate to construction-site router work is the bearing-guided flush-trimmer router bit (photos below). Because the guide bearing is the same diameter as the cutter, there is no need to account for the offsets typical in base-guided and collar-guided setups. This makes it easy to whip together perfect jigs.

Flush trimmers are available in several configurations. A bearing-over bit has the bearing mounted on the shank-end of the bit (photo below). Its cutter is retained by a nut at the end of the shank. A bearing-under bit has the bearing at the other end. I often use this kind of bit in my router table.

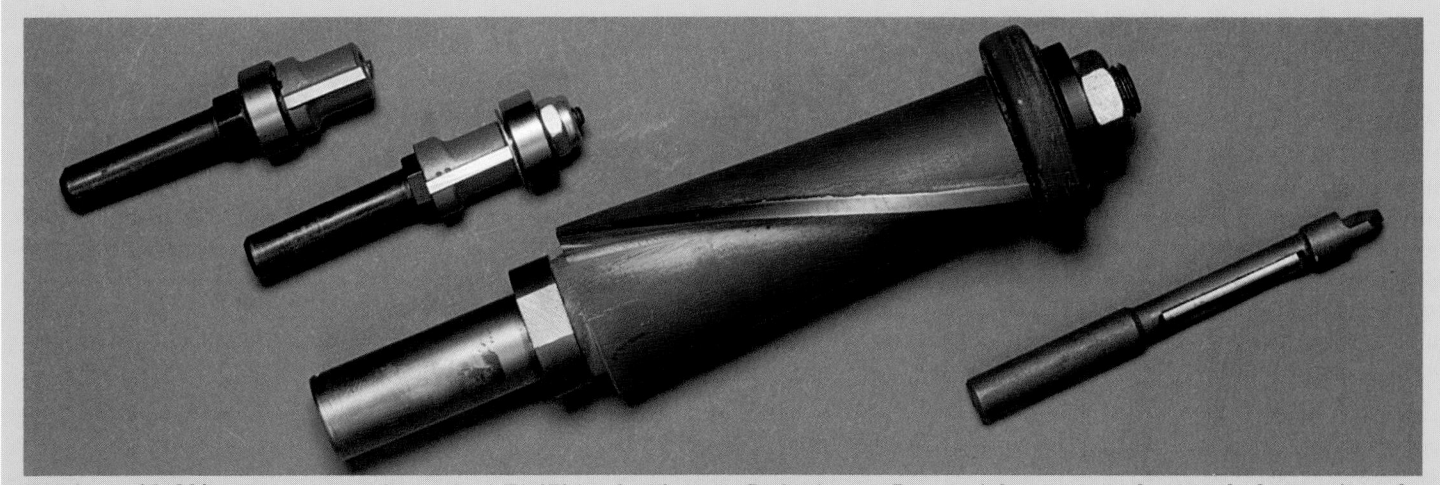

Bearing-guided bits. **Shown from left to right: A TA-170 is a *bearing-over* flush trimmer. Because it has a recessed nut on the bottom, it can be used to cut mortises. A *bearing-under* bit can be useful mounted upside down in a router table. A *face-frame* bit, called an FFT-2126, is a bearing-under bit with 2½-in. long spiral cutters. The pointed end of the *panel* bit allows it to be plunged through the workpiece.**

From *Fine Homebuilding* magazine (August 1989) 55:36-41

The flush-trimming bit I use for template mortising is called a dado bit and is made by by Paso Robles Carbide (731 C Paso Robles St., Paso Robles, Calif. 93446). It is called the TA-170, and it has a recessed nut on the bottom, allowing it to cut mortises. A type of flush-trimming bit that I use occasionally for cutting shapes in thin material (¼ in. or less) is called a panel bit (photo facing page). This inexpensive bit is made of high-speed steel, and it has no roller bearing. But its point allows the bit to be plunged through the workpiece. The bit also has a bearing surface that can be used like a bearing-under bit so long as the pattern against which it is riding is absolutely void-free and accepting of a little friction burning.

Large-diameter bits cut much better than small-diameter bits. If you plan on removing a lot of wood, as when mortising housed stringers, you'll be better off with a ¾-in. bit. For hinge and strike-plate mortises, a ½-in. bit is fine.

Most flush-trimming bits are available with spiral cutters. Of these there are two varieties: upshear and downshear. Upshear bits keep the material snug to the base of the router or to the router table. They also draw the sawdust out of mortises. Downshear bits aren't useful for the applications discussed here.

I see no point in buying a router bit made of any material inferior to tungsten carbide. The demands on the cutting edge of a bit are too great for anything less. Sharpening will still be necessary, though, and this presents a problem for flush trimmers. Sharpening reduces the diameter of the cutter, making it no longer flush with the bearing. With small-diameter trimmers the solution is to use detachable throw-away cutters (such as the TA-170) that allow the arbor and bearing to be saved. With large-diameter trimmers you can get fiber-covered bearings that can be resized to match the new diameter of a resharpened cutter.

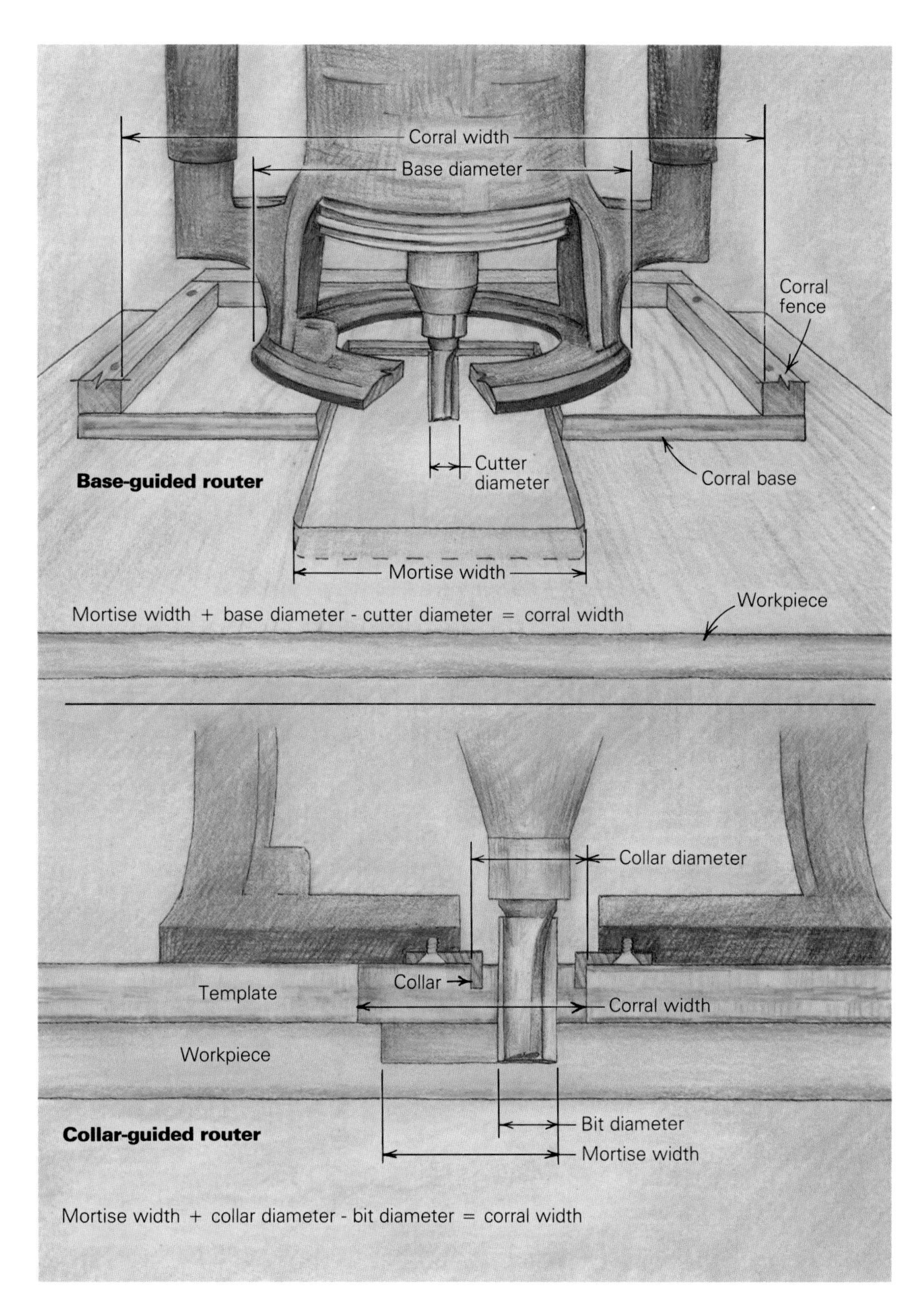

Hinge butt templates—Store-bought hinge butt templates (see *FHB* #31, pp. 28-31) rely on a collar-guided router, and this is their main drawback. The bit, collar and template must all be compatible for the setup to work. Also, the collar has to match up with the router you plan to use. And once you've got all these things sorted out, if you hit the steel template with your bit, you've ruined both of them. As an alternative, I devised the hinge-mortising jig shown in the drawing (bottom left drawing, next page). I begin making one of these jigs with a piece of void-free material such as high-quality birch plywood or medium-density fiberboard (MDF). I pull the pin of the hinge I intend to use on the door and place the unfolded hinge, barrel up, on the template stock (top left photo, next page). Now I corral the hinge with fences that are thin enough to allow the bit's cutters a ½ in. or so of entry into the plywood as the bearing rides against the strips. The height and width of this corral is determined by the unfolded size of the hinge. You don't even have to measure the length of the strips because they can run wild at the corners.

Routers are noisy and they scatter a lot of wood chips. Before you make any cuts, protect your eyes with a pair of goggles that allow you a good view of the work, and use either foam plugs or the earphone-type headgear to protect your ears.

Using a dado bit, rout clockwise along the inside edges of the corral. If you go the opposite direction, the bit will try to pull its way into the wood as you advance the router, making it difficult to control.

Chances are, your bit didn't go all the way through the template stock because the cutters aren't long enough. Remove the wood strips and use the walls of the grooves you've already cut to guide the bit's bearing as you complete the cut with another pass. Consider this template the master copy, and use it to cut three working templates. That way, if you damage a working template you can use the master to make a new one in minutes, without even changing the bit. I make the working templates for hinge mortising out of ¾-in. stock, which is thick enough to allow the bearing to ride against the edge of the template while the bit cuts a shallow mortise (top right drawing, next page). This principle holds true for any template—it has to be thick enough to guide the bearing, yet thin enough to allow the cutters into the workpiece. Figure on a ½-in. deep cut as the maximum that you'll be making in one pass.

The rounded inside corners of templates cut with a ½-in. bit match the radius of hinges with rounded corners. If your hinges have square corners, use a chisel to square the in-

Drawings: Chris Clapp

To make a hinge template, unfold the hinge (minus the pin) and corral it with wood strips to guide the router-bit bearing.

The author uses his hinge-mortising jig to simultaneously cut hinge gains in the door stile and the jamb.

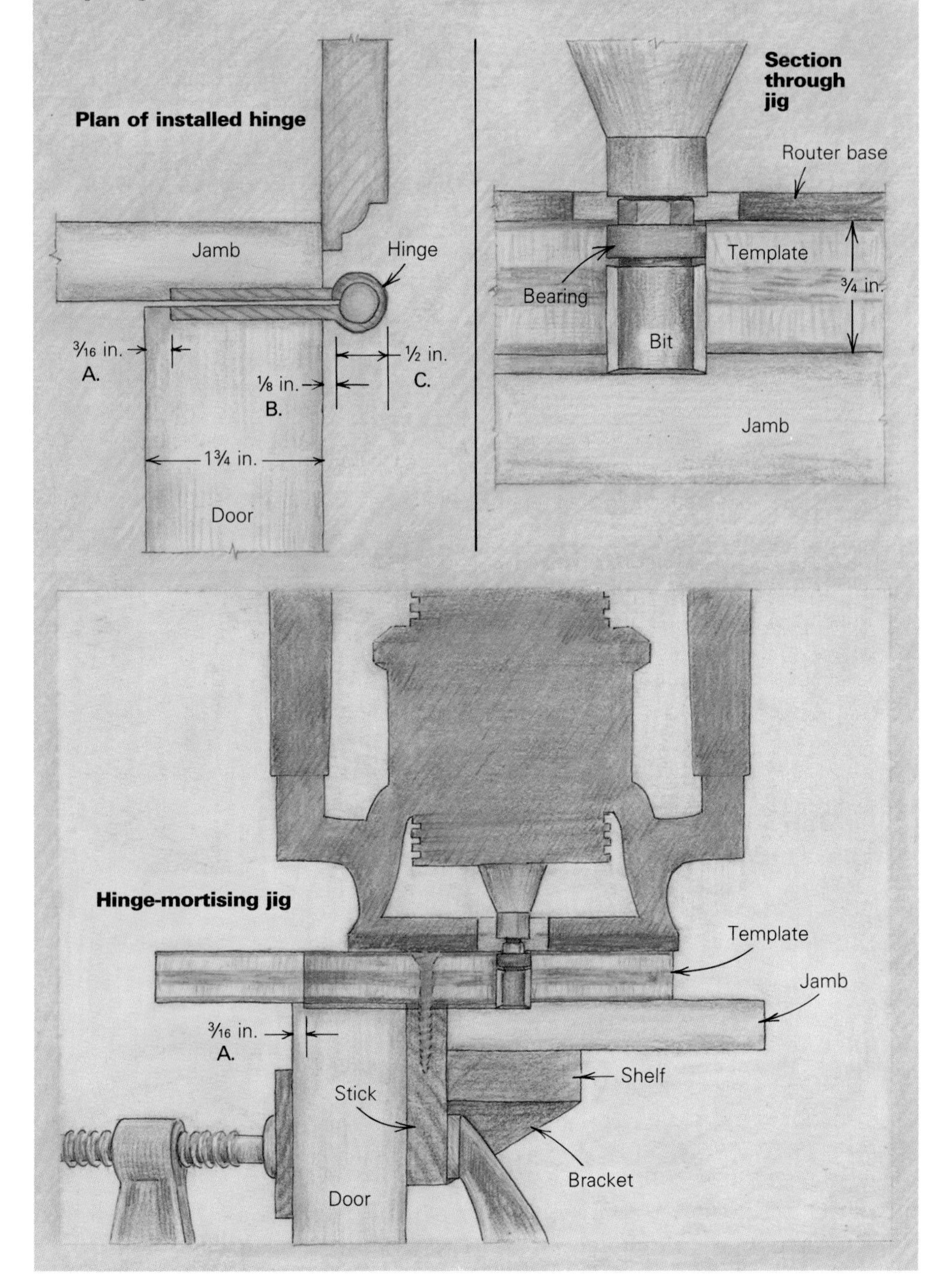

side corners of the templates. The squared walls will act as a guide for the chisel as you square the corners of the hinge gains.

Making a jig from the templates—I screw the templates to a stick that fits between the door and the jamb (drawing below left). The stick's thickness is determined by the distance that the hinge will protrude beyond the plane of the door. For example, the hinge on the 1¾-in. thick door shown in the drawing is held back 3/16 in. from the edge of the door at A. The leaf of the hinge projects ⅛ in. past the door at B, and the barrel is ½ in. in diameter at C. The stick thickness is equal to the diameter of the hinge barrel plus two times B. That adds up to ¾ in. in this case. Three brackets mounted to the side of the stick support a shelf upon which the jamb rests. In use, a clamp at each end secures the jig to the door (top right photo).

The jig can be made reversible for left-hand or right-hand doors in two ways. First, you can make the jamb-shelf easily removable so that the position of the jamb and door can be switched. Another way is to make the projection of the stick past the top and bottom hinge template equal, so that the whole jig can be switched end for end. In either case make the stick ⅛ in. longer than the door to allow for clearance at the head jamb.

Stair-mortising jig—Currently on the market is a metal jig for cutting housed stair stringers with a router. Because it's made of metal, it can damage your bits, and it costs a fair amount of money. In the time it takes to go to a nearby store to buy one, you can make a better one of your own.

On a piece of voidless plywood approximately 24 in. by 30 in., drill a hole the same diameter as the thickness of your tread stock. This hole corresponds to the tread's bullnose, and it should be in roughly the position shown in the top drawing, facing page. Tangential to this hole draw a line parallel to the long edge of the plywood. This line represents the top of the tread. Draw a second line to represent the front of the riser; it should be 1 in. to 1½ in. back from the farthest reach of the hole. Carefully square this line to the first one. Tack wood strips along these lines, followed by some offcuts of your tread and riser material, and some samples of the wedges you intend to use (drawings, facing page). Corral this assembly with more strips of wood, remove the tread, riser and wedge samples and use a dado bit to rout a master template. Make a couple of working templates from the master.

At this point the housing template you've made is good for any rise and run of stair using your tread and riser stock. It has no pitch—only the rise and run. In this configuration you can use the template like a framing square, measuring the rise along the face of the riser mortise, and the run along the top of the tread mortise.

I use a pair of these templates to create a jig that can be used to rout the mortises in a pair of fully housed stringers (bottom drawing, right). I screw the two templates together with some 1x2 braces, making sure to keep them far enough from the mortises to allow the router clearance. Fences on the underside of the assembly align it with the stringers.

After you've routed the first pair of stair mortises, you can affix a pair of blocks to the underside of the jig. They are positioned to fit into the previously routed tread mortises, allowing you to advance the jig accurately without measuring.

Table-saw router table—Negative shapes, such as mortises, are best cut with negative templates and a dado bit. I cut positive shapes, however, on a router table using a pattern and a flush-trim bearing-under bit. My router table is actually an extension wing that I affixed to the side of my table saw by way of bolts run through threaded holes in the saw's guide rails and cast-iron table. This arrangement saves space in my garage/shop.

I made my extension wing out of Baltic birch plywood, and covered it with plastic laminate. It's easy to keep clean and materials slide across it with little effort, but to tell the truth, next time I'll omit the laminate. A sanded plywood surface is smooth enough, and I could easily attach fingerboards, fences and guards to the plywood with drywall screws without worrying about marring its finish.

Router-table safety—The safest way to use the router table with a fence is to make the cut so that direction of feed opposes the spin of the cutter (bottom left drawing, next page). A fingerboard guarantees that the material doesn't drift away from the cutter, resulting in a tooled edge with bulges and ripples. In my setup the workpiece passes to the left of the cutter. If I moved the fence over so that the workpiece could pass to the right of the cutter, I'd have a setup for sending wooden projectiles through my shop walls at high velocity.

A typical freehand router-table setup is shown in the drawing and photo top right, p. 83. The spin of the cutter is countered by the direction of feed, the guide bearing and the starting pin. A starting pin is simply a post or even a small block of wood that braces the workpiece as it enters the cutters (but before the stock has reached the guide bearing). It can keep the whirring bit from ripping the workpiece right out of your hands. My starting pin is an Allen-head screw driven into a teenut let into the underside of the table. To make it friction-free, I clad the exposed portion of the screw's threads with a short piece of chromed-brass plumbing-supply tube. Epoxy holds the assembly together.

The work has to go counter to the rotation of the bit. To make sure that I don't forget its direction, especially during repetitive production operations, I use a felt-tip marker to draw the cutter's rotation on my table.

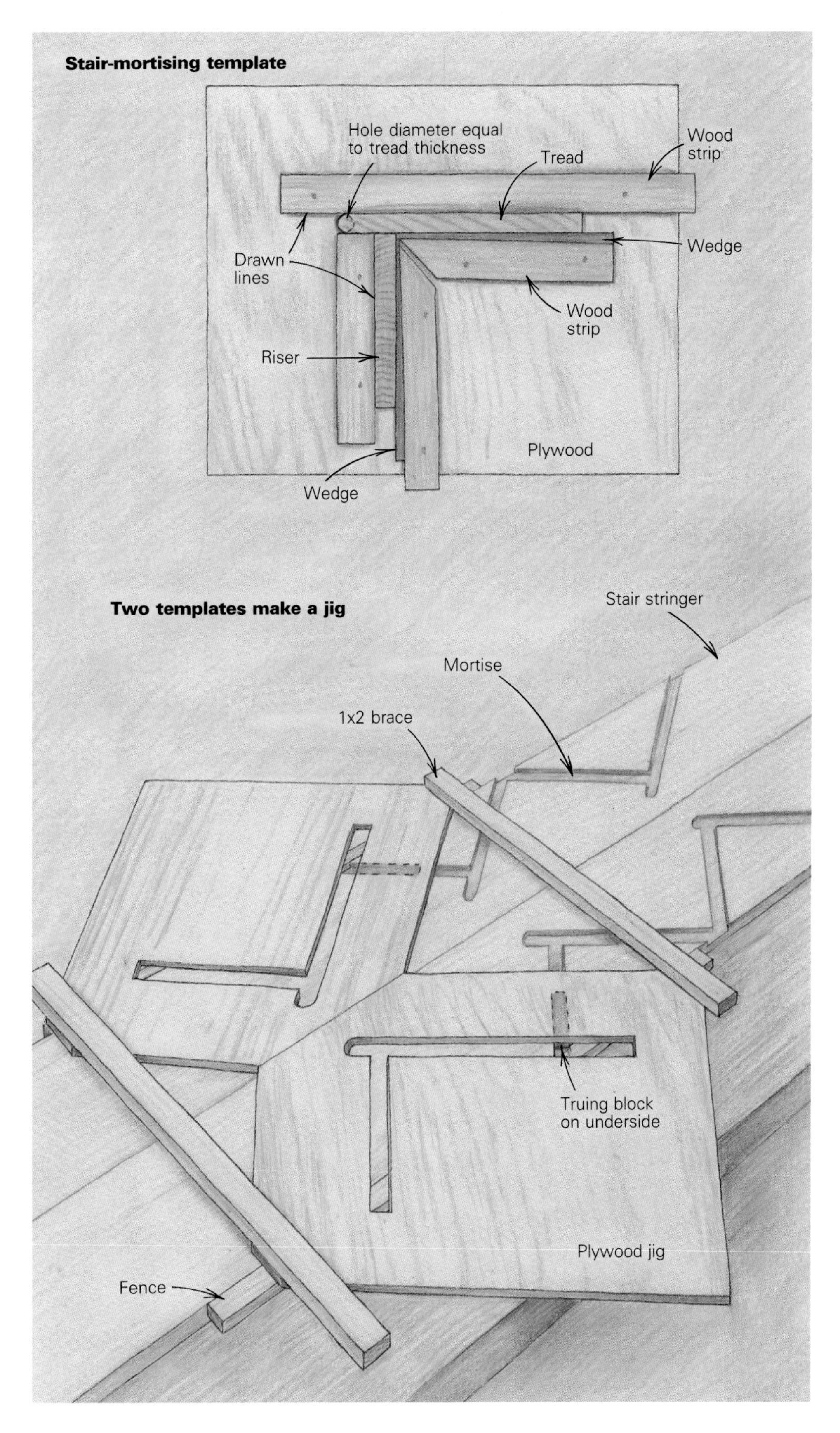

There is one setup for the router table that allows stock to be fed *with*, rather than against, the spin of the cutter. The drawing (bottom right drawing, next page) shows how this is done. This is called climb-cutting, and it's for dealing with heavy pieces of difficult woods. To make it safe, you must have a secure table and take a light cut (1/16 in. or so) off a fairly massive piece of wood. Milling the bullnose roundover on Douglas fir stair treads is a typical application of this technique. The stringy grain of the wood makes this backwards approach necessary, because the conventional method would tear the grain. In climb-cutting, a heavy workpiece that is easy to grip is essential for control.

Mounting the router—My R-330 Ryobi router has two large holes for its removable handles. Two studs that I permanently af-

Peake affixes his router to a table-saw extension wing with a pair of bolts through the handle-brackets. Note the L-shaped flange screwed to the side of the router. The orange depth-adjustment ring bears against it in this inverted position, allowing the bit to be easily raised and lowered. A T-nut is embedded in the underside of the table near the router's base.

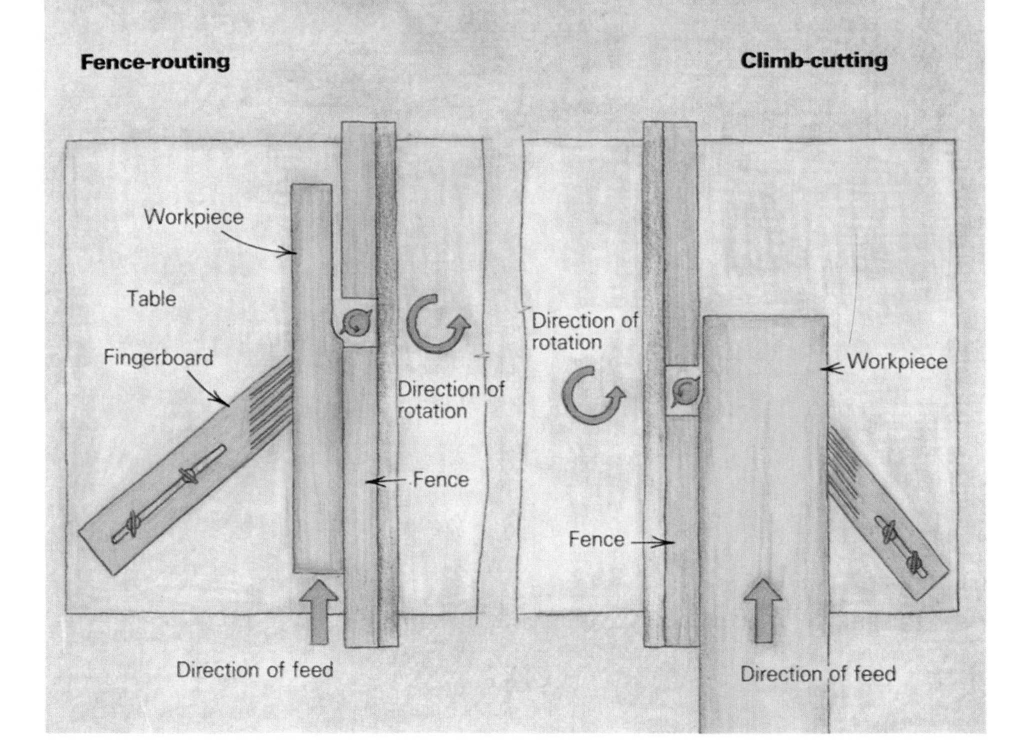

fixed to the underside of the table pass through these holes, allowing the router to be held firmly in place with just two wing nuts (photo left).

A router that lacks these bolt holes can be secured by running screws through the table into holes tapped in the base of the router. If these screw holes are metric, it's usually easier to redrill and retap them with English threads rather than try to find the right metric screws.

I've yet to see a router with a depth adjustment that is easy to use when the router is upside down. This is important enough in table-routing that I added an L-shaped metal flange to the side of my router to give the depth-gauge threads something to bear against (photo left).

It's awkward to fumble around under a table, feeling for an unseen switch to turn off a screaming router. To avoid this scene, I mounted a switched outlet to my table saw's base. It's in an accessible position so I can easily flick the tool on or off.

Pattern-routing—I worked on a Victorian hotel recently that has fancy redwood wainscoting bordered by custom trim. The vertical trim pieces framing each wainscot panel have two curves and a straight section, and I used the router table to mill them quickly and accurately. I started by making a pattern out of voidless ⅛-in. plywood that represented the profile of the trim piece. Then I used a bandsaw to stack-cut a bunch of redwood trim blanks that were about ¼ in. oversize.

To mill each trim blank down to its finished size, I tacked the plywood pattern to the top of a blank and ran the pattern and blank against a flush-trimming bit (pattern-routing photo, facing page). To make sure the pattern made good contact with the bit's bearing, I used thin plywood spacers between the pattern and the blank.

Once I had all of the pieces trimmed to their finished size, I put an ogee bit with a pilot bearing in the router and ran a decorative edge on the trim using the same setup without the pattern. Because the radius of the router bit cannot make the crisp interior corners appropriate to some parts, I used a sharp gouge to clean up the inside corners on these trim pieces.

Jointing on a router table—With some additions to the fence that came with my table saw, I can use the router table to dress the edges of boards with jointer-like accuracy. I wrap the fence with ½-in. Baltic birch plywood. This gives me a surface to which I can affix wood blocks that create a space for the router bit. Rather than use blocks of the same thickness on both sides of the bit (which would work fine for most fence-based routing operations), I made the infeed side adjustable. By aligning the outfeed side of the fence flush with the cutter and offsetting the infeed side, I control the depth of cut. To

Adjustable infeed fence. **Oak blocks with dovetailed ways compose the infeed fence on Peake's router table (photos above and below). A barrel nut in the middle block is tapped to accept threaded rod (drawing below). Turning the handle clockwise moves the outer blocks away from the fence; counterclockwise retracts the blocks as a collar affixed to the threaded rod bears against a mortise shoulder in the block nearest the handle.**

¼-in. threaded rod

Barrel nut

Collar

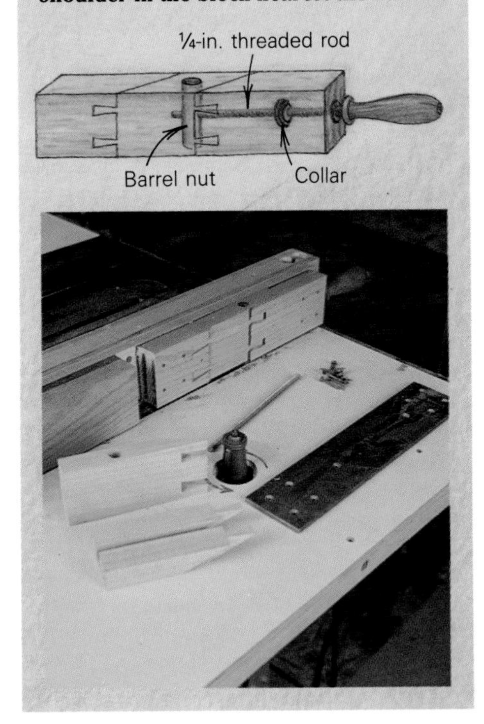

Pattern-routing with flush-trimming bit

Bearing

Pattern

Bit

Workpiece

Router table

Pattern-routing. **Precise, repetitive shapes can be quickly milled on a router table by tacking a pattern to the top of a blank and then passing the assembly against a flush-trimming bit with a bearing on the cutter end of the shank. To make sure that the pattern fully engages the bearing, a spacer can be placed between the workpiece and the pattern, as show in the drawing above. An arrow drawn on the table indicates the spin of the bit.**

Indexing. **Peake uses a sliding extension table on his table saw to cut dovetails on the end of drawer stock. Successive cuts are indexed by the aluminum registration pin.**

make the infeed fence adjustable, I cut dovetails in three 4/4 oak blocks. The center block is affixed to the fence, and it has a vertical pin in it with a threaded hole (adjustable infeed-fence drawing and photos above). The other two blocks nest on either side of the fixed block and are connected by a ¼-in. steel plate. A threaded rod passes through a lengthwise slot in the outboard block and into the pin. By turning the threaded rod, I expand or contract the thickness of the infeed fence.

The bit I use for this is called a "face-frame" bit (photo, p. 78), and it has 2½-in. long spiral cutters. Paso Robles Carbide is the only company I know of that makes it. Because the bit is so long, I can raise its bearing above the top of the typical workpiece during jointing operations (photo below right).

Whenever I'm using the fence to guide long, thin stock past a profiling cutter head, I use a fingerboard to apply pressure from the side. To make it adjustable, I put a couple of T-nuts in strategic places under the router table. A slot in the fingerboard and some wing nuts let me tighten it down where necessary.

Indexing—I made a sliding table for my table saw (for more on sliding tables see *FHB* #53, pp. 58-61). It has two hardwood runners that slide in the table-saw guide slots. These runners are attached to the underside of a piece of ½ in. birch plywood that is nearly as large as the table-saw top. Topside at the front and rear of this piece of plywood, perpendicular to both the runners and the saw blade, are wooden fences. I use the sliding table to crosscut material with the table saw, but I also use it for index cuts with the router. Indexing is a clever but simple way to produce repetitive, equally spaced cuts such as dovetails. Start by making a ⅛-in. thick aluminum registration pin that matches the profile of your dovetail router bit (indexing photo above). This pin must be set away from the path of the dovetail bit by an amount equal to the bit's profile (this will take some trial-and-error fussing). Once you've got it close, you can adjust the final spacing of the joint by slightly raising or lowering the bit.

Rout the sockets of the dovetail with the stock flat on the sliding table. You can affix blocks to the router table to act as depth stops. The tails are routed by standing the stock on end, flat against the back fence of the table. I use a 1x2 screwed to one end of the sliding table as a clamp to secure the stock while it passes over the bit. The 1x2 also serves to back up the cut, preventing tearout.

With some layout work you can calculate where to start the first pin and first tail so that the edges of the dovetailed boards align at the joint. To keep it simple, I work with extra-wide stock and rip the whole carcase down to size after the joint has been assembled. □

Jud Peake is a contractor from Oakland, Calif., and a member of Carpenter's Local #714.

Clamps on Site

Select the right tool to get a grip on your work

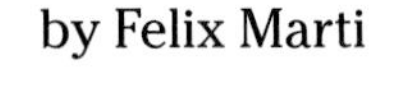

by Felix Marti

Clamps are like thumbs—they're easy to take for granted and hard to live without. But unlike thumbs, you can't have too many of them. In this article, I'll talk about the array of clamps I use around the job site (along with some of the newer, specialized clamps), and I'll offer tips on how to get the most out of them.

Most of the clamps I use are available from any well-stocked hardware store or mail-order tool outlet. A few can be hard to track down, however, and the "Sources of Supply" on page 87 will help you get in touch with manufacturers. If you can't find what you're looking for commercially, consider making your own. The ability to modify or make a clamp of some sort to solve a specific problem is a hallmark of the ingenious builder.

C-clamps—About as basic as a clamp can be, the C-clamp is the workhorse around our jobs. It consists of a C-shaped frame with a machine screw at one end (photo below right). A swiveling tip at the end of the screw bears against a foot at the opposite end of the frame. C-clamps are measured by how far they will open, and by their throat depth or "reach"—the distance from the inner edge of the frame to the center of the screw. You should use a clamp that opens just a bit more than the thickness of the material to be clamped; otherwise, you'll place abnormal stresses on the screw and the frame. For site work, I find that 3-in., 4-in. and 6-in. clamps fit most situations. I organize the clamps according to size, store them in joint-compound buckets and lubricate their threads occasionally.

Pipe clamp. **A sliding foot, or tail-stop, can be moved up and down a length of ¾-in. pipe to adjust the distance between the clamping surfaces. Here, a pipe clamp aligns a pair of rafters while another brings the rafter and the subfascia into the same plane. Photo by Felix Marti.**

C-clamps are designed to be hand-tightened. If you haul on one with a pipe lever or a wrench, you'll likely ruin it by deforming the frame. C-clamps are also designed to clamp parallel surfaces. If you have to clamp two surfaces that aren't parallel, use wedges or angled seat blocks to direct the clamping force to the foot. This will keep the screw from bending.

Some clamps have plastic pads over the swivels and feet to protect the work, but I take them off when using the clamps for glue up. They're squirrelly, tending to move around a bit as the clamps are tightened, and they don't do anything to spread clamping pressure the way a caul will. My favorite cauls are strips ripped from sink cutouts. They have laminate on one side, which won't bond with the glues I use, and their ¾-in. thickness spreads the pressure well. For thin cauls, I use ¼-in. hardboard.

Two clamps. **A quick-action clamp on the left and a C-clamp on the right secure a jamb extension to a 90° angle jig.**

To determine quickly and accurately the distance between two surfaces, I'll often use a pair of long 1x2s and a C-clamp (two C-clamps if the sticks are spanning more than 8 ft.). I slide the 1x2s apart on their long axis until they abut the opposing surfaces. Then I clamp them together to record the exact distance without a tape measure.

C-clamps also make good handles for small, hard-to-grip loads on the job site. During remodeling, for example, you can use one to get a firm hold on a piece of blocking that must be carefully removed from a joist or a stud bay.

Pipe clamps—Pipe clamps consist of two elements affixed to a length of pipe: a head that contains a screw and the crank; and the foot, or tail-stop, that carries the anvil pad against which the pressure is applied. Together, the head and foot are called the *fixture*. One element of the fixture is stationary; the other slides up and down the pipe as needed and is held in place by a spring-action clutch. I prefer a pipe clamp with a stationary head and a sliding foot because it allows room for a big crank handle that can be easily turned. Sliding heads are fitted with wing-nut handles because there isn't much clearance between the head and the pipe. Pipe-clamp fixtures are made for both ½-in. and ¾-in. pipe. Stick with the ¾-in. stuff—it's more stout. And be sure to use black pipe, not galvanized. Galvanizing hardens the pipe surface, making it difficult for the sliding clutch to get a bite. I have pipe clamps in 2-ft., 4-ft. and 6-ft. lengths, and I typically keep couplings on their ends to protect the

From *Fine Homebuilding* magazine (April 1992) 74:70-73

threads and to allow the clamps to be linked into longer clamps.

When I fasten subfascia to rafter tails, I align the top of the subfascia with the tops of the rafters. Rafters aren't always straight, so this alignment sometimes requires a fair amount of persuasion—perfect for a pipe clamp (top photo, facing page). Pipe clamps often come loose when vibrated, so a nail gun or a screw gun is good for making this connection.

When building cabinets, I run pipe clamps diagonally across casework to square the carcase prior to applying the backing panel. I also square frame-and-panel doors with pipe clamps before C-clamping the corner joints during glue up.

Bar clamps—Bar clamps are a variation on the pipe-clamp design, but instead of a pipe for a frame, they use a solid-steel rail or a square metal tube. Like pipe clamps, bar clamps are often used to glue boards edge-to-edge, but both tend to bow the boards away from the clamps. You can counteract this phenomenon by using cauls and clamps placed in opposition to the bowing, or with edge clamps (top photo, right).

Truth be told, I don't care much for traditional bar clamps. Bar clamps that are as sturdy as ¾-in. pipe clamps usually cost more than pipe clamps, and they can't be extended by linking them together. But there are some new versions of bar clamps on the market that are worth noting. For example, Bessey's K-body bar clamp has 3½-in. deep jaws that remain parallel throughout their clamping range. According to the Bessey folks, steel rollers within the sliding jaw evenly distribute the clamping pressure, eliminating the bowed-board phenomenon. To verify their claims, I ripped some ½-in. plywood into nine strips, 2½ in. wide. Then I placed the strips edge-to-edge and drew them together with a pair of K-body clamps. Nothing bowed or popped.

Universal's aluminum bar clamps are well made, sturdy and very light, which can be a real plus when it comes to moving a glued-up assembly with lots of clamps on it. In addition, they can quickly be converted to miter clamps with the addition of slip-on miter jaws (photo right).

Both Adwood and Häfele make bar clamps that have wide, cylindrical steel jaws wrapped with PVC plastic sleeves (photo left). The wide jaws exert pressure over a broad area, and their cylindrical profile allows them to draw together workpieces that meet at other than 90° angles—for example, the carcase of a triangular cabinet.

The Quick Grip clamp, on the other hand, is a new type of bar clamp that, in my opinion, has limited use on the job site. It's a good idea: a squeeze-grip handle allows you to exert the clamping pressure one-handedly. Problem is, the pressure is minimal, and I don't like the handle.

Edge clamp in the field. **Affixed to a bar clamp, an edge clamp can exert downward pressure to counteract the tendency of edge-glued boards to bow upward.**

Miter fixture. **Universal's lightweight aluminum bar clamp (photo below) can be converted into a miter clamp with the addition of slip-on, V-shaped jaws.**

Cylindrical jaws. **Häfele's top-spanner bar clamp (below) has PVC-wrapped cylindrical jaws that can get a bite on workpieces that come together at any angle between 45° and 135°. Photo courtesy Häfele.**

Quick-action clamps—Cross a bar clamp with a C-clamp and you've got a quick-action clamp (bottom photo). It has a sliding head that can be moved along the bar for fast contact with the work. This feature is especially helpful when using the clamp to align heavy or unwieldy parts, such as framing members. Quick-action clamps can have deeper reaches than C-clamps (up to 9 in.) and can be opened wider—Jorgensen quick-action clamps are up to 60 in. long. On site, I keep several with 4-in. reaches and 8-in. to 12-in. bars and some with 8-in. reaches and 16-in. bars.

I often use screws, counterbored and plugged, to affix door casings to framing. A couple of 12-in. quick-action clamps do a good job of holding the casing in place and snugging it to the drywall at the same time (softwood cauls protect the casing from damage).

When assembling window-jamb extensions, door jambs or other butted joints, I use C-clamps or quick-action clamps along with a simple 90° angle jig to align and hold the pieces while I screw them together (bottom photo, facing page). Another jig that can be used with C-clamps or quick-action clamps makes it easy to bend sheet metal on site (drawing, p. 87).

Part of what makes a quick-action clamp useful can also render it useless. If the screw is fully retracted and you slide the head down the bar to engage the foot, that's where it will stay, wedged shut. To prevent this, always keep a few threads exposed between the head and the swivel tip for backing-off room.

Locking clamps—Anyone with a passing interest in hand tools has come across locking pliers—the

Beyond the C-clamp. **Locking clamps, pistol-grip bar clamps and quick-action lever clamps exert similar loads as C-clamps do but do it faster and one-handedly. Left to right: American Tool Co.'s bar clamp has a sliding jaw with a locking lever, and its C-clamps have a wide range of reaches. Center: a Quick Grip bar clamp. Last three: Bessey's all-steel quick-action clamp, lever clamp and Super Grip clamp. Photo by Rick Allen.**

Hold-down. **De-Sta-Co clamps apply a point load by way of a lever-activated cam.**

Spring clamp at the corners. **A pivoting-jaw spring clamp has tiny teeth that secure mitered corners, 90° or otherwise.**

Edge clamping. **Combined with a quick-action clamp, Wetzler's edge screw applies pressure to a counter edge strip (left). Adwood's heavy-duty edge clamp has self-adjusting, spring-loaded jaws (right).**

most common brand is the Vise Grip. Their close cousins are locking clamps, and they can be very useful on the job site for applying pressure one handedly, even in situations that call for wide openings or deep reaches (bottom photo, previous page). I use them for jobs like gluing veneer to a substrate, or plastic laminate to countertops.

De-Sta-Co makes a lot of cam-over clamps that fall into the locking-clamp category. You adjust the jaws to engage material tightly when you throw the lever handle. I use their fixed-mount clamps on sliding table-saw fixtures or to hold a workpiece during routing (top photo, left).

Bessey's lever clamp combines a locking-lever jaw with the rail of their all-steel, quick-action clamp (bottom photo, previous page). The result is a heavy-duty clamp that can exert 50% more pressure than their comparable quick-action, all-steel clamp.

Handscrews—Handscrews are my clamps of choice around the table saw—I don't like steel near spinning saw blades. I use handscrews whenever I have to clamp a fixture to the fence or a featherboard to the table. I also use a handscrew as a stop on my radial-arm saw's fence. I turn it at an angle toward the blade so the workpiece engages the corner of the handscrew's jaw, preventing sawdust buildup against the stop.

Clamped to the bottom of a door, a handscrew can make a serviceable door buck—it's usually wide enough to keep the door from falling. Clamping a handscrew to each end of a board, then clamping the handscrews to a pair of sawhorses creates a pretty good field vise for boards requiring edge treatment, such as planing or sanding (photo top right).

Because a handscrew has two screws, its jaws can be perpendicular or at an angle to each other. This characteristic allows a handscrew to conform to odd shapes or to pinch nearly inaccessible places with a point load. If you've got some threaded rod and a couple of 2x4s, you can make a site-built version to reach tough spots (drawing, p. 87).

Spring clamps—Spring clamps are light-duty clamps that resemble big clothespins. They don't exert much pressure, but they can be operated with one hand, and they're inexpensive. I use spring clamps for holding steel studs in their channels prior to screwing the studs in place. They are also good for holding a chalkline, anchoring a tarp where tape won't work or securing a template to a workpiece.

The spring clamp I wouldn't be without is the pivoting-jaw miter clamp (middle photo, left). Its jaws have tiny teeth that grip opposing pieces of wood. The clamps make it easy to assemble runs of baseboard or crown with glue and air-driven nails—especially a run with short pieces in it.

Strap clamps—Once in a while, I'll have to bring together a number of parts that can't be easily gripped with ordinary clamps—cylindrical assemblies like hot tubs and planters, for example. They are best held together temporarily with strap, web or band clamps. This type of clamp encircles its subject with a belt that can be tight-

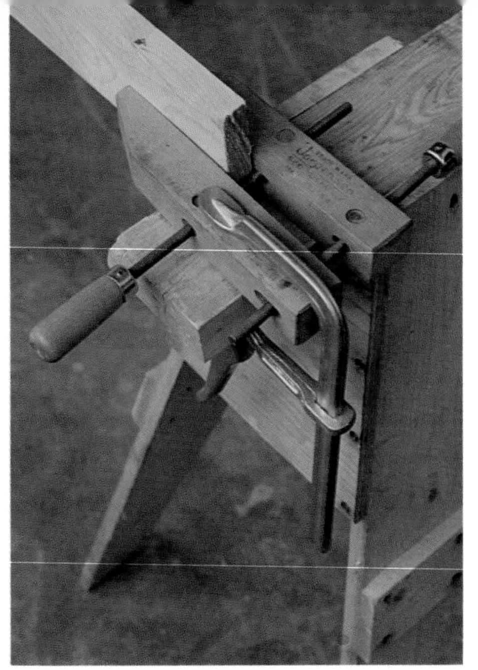

Clamp-on sawhorse vises. **A sturdy anchor at one end of the sawhorse can make it a lot easier to plane the edge of a board. Stanley's steel vise (photo below) has prebored holes for attaching wooden jaw liners. Bessey's version has plastic jaws that ride on a pair of guide bars (bottom photo). The bars can be removed, allowing the jaws to swivel and thereby accommodate irregular shapes. A handscrew clamped to the end of a sawhorse can also make a serviceable vise (photo above).**

ened with a self-locking cam device. Band clamps usually have four preformed metal corners that can be threaded onto the strap and used to draw mitered frames together.

Two less sophisticated (but cheaper) ways to apply similar clamping action are to use rope or inner tubes. For example, tie a loose noose of rope around the workpiece, pass a sturdy stick through the noose and twist the stick until it tightens the noose. Anchor the stick with a block or a nail. I've also used loops of truck inner tube cut into big rubber bands to make clamps for gluing up columns. They can be linked to make any length.

Wedge clamps—When I lay flooring, I snug bowed strips together using a pair of wedges cut from a 1-ft. long 1x6 (drawing below right). The fixed wedge is screwed to the subfloor to act as an anvil. The driving wedge has a grooved edge that bears against the tongue of the flooring. I lop off the point of this wedge to create a flat spot to knock it back out.

I also use wedges to plumb and line framed walls. I put pairs of opposing wedges between the brace-blocks nailed to the subfloor and the bottoms of the 2x4 braces. Moving the wedges in and out allows me to fine-tune the alignment of the walls.

Irregular angle clamp. **Made by Bessey, this two-jawed clamp brings together parts that meet at odd angles, such as this stair railing. Photo by Felix Marti.**

Clamp chowder—In addition to the basics, many specialty clamps on the market can take the pressure off the builder and put it on the workpiece, where it belongs.

I have a couple of clamp-on carpenter's vises that are made specifically for sawhorses (photos facing page), and I'm surprised at how much I use them. Their jaws open up to 3½ in.—enough to do edge work on most job-site materials. Stanley and the Warren Tool Group both sell them. Bessey's version rides on removable rails, allowing it to clamp irregularly shaped workpieces.

When you have to apply edging to a counter or a cabinet, and there's no easy purchase for a pipe clamp, edge clamps can bail you out (bottom left photo, facing page). For occasional use, I'd suggest edge screws, which are used in conjunction with a quick-action clamp to bear on the edging.

Adwood sells the ultimate edge clamp. Based on the old Chinese woven finger trap that tightens as you pull on it, Adwood's heavy-duty edge clamp automatically adjusts to accommodate panels from ⅜ in. to 3⅛ in. thick. But at around $90 apiece, it might be hard to justify their price tag.

Bessey's irregular angle clamp is another tool you don't really need until you really need it (photo above). It seems best suited for joining stair parts and similar assemblies that meet at angles other than 90°. I tried one out as I researched this article and ended up buying it. □

Felix Marti is a builder in Ridgway, Colo. Photos by Charles Miller, except where noted.

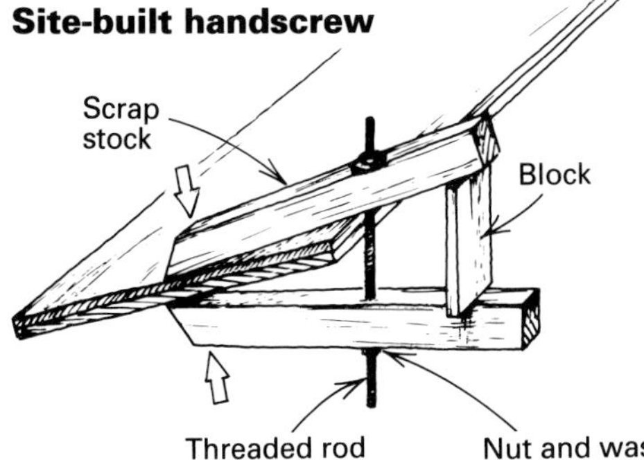

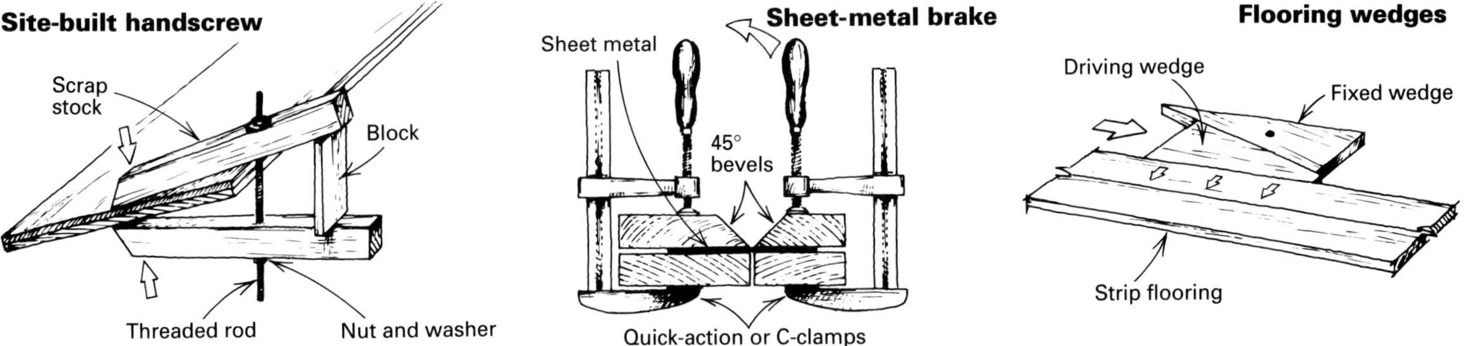

Sources of Supply

•Adjustable Clamp Co., **404 North Armour St., Chicago, Ill. 60622 (312) 666-0640**
Makers of Jorgensen & Pony C-clamps, pipe, bar, quick-action, spring and band clamps; handscrews and hold-downs.

•Adwood Corp., P. O. Box 1195, High Point, N. C. 27261 (919) 884-1846
Wide, cylindrical-jawed bar clamps for cabinets, heavy-duty edge clamps.

•American Clamping Corp., P. O. Box 399, Batavia, N. Y. 14021 (716) 344-1160
Imports Bessey clamps—the most innovative clamps around. They include angle and miter clamps, quick-action, Klemmy cam-style veneering clamps. Also locking clamps, bar clamps and K-body bar clamps.

•American Tool Co., Inc., 108 So. Pear St., DeWitt, Neb. 68341 (402) 683-2315
Vise-Grips; C-clamps with swivel pads and regular tips, quick-set bar-clamp jaws, sheet metal and locking pipe jaws, chain jaws.

•De-Sta-Co, P. O. Box 2800, Troy, Mich. 48007 (313) 589-2008
Special-purpose clamps: cam-over, pull-action, push-action, air and hydraulic powered, bar clamps, wrench-tightened clamps and squeeze clamps.

•Häfele American Co., 3901 Cheyenne Dr., P. O. Box 4000, Archdale, N. C. 27263 (919) 889-2322
Top-spanner cabinet clamp.

•Stanley Tools, The Stanley Tool Works, 600 Myrtle St., New Britain, Conn. 06050 (203) 225-5111
C-clamps, bar clamps, carpenter's vises, web clamps.

•Universal Clamp Corp., 15200 Stagg St., Unit #3, Van Nuys, Calif. 91405 (818) 780-1015
Lightweight, sturdy aluminum clamps and face-frame clamps.

•Warren Tool Group, P. O. Box 68, Hiram, Ohio 44234 (800)-543-3224
Brink and Cotton line of C-clamps, pipe, bar, quick-action, spring and web clamps. Carpenter's vises and handscrews.

•Wetzler Clamp Co., Inc., Rt. 611, Mt. Bethel, Pa. 18343 (717) 897-7101
Pipe, bar, quick-action, spring and band clamps.

•Wilton Corp., 300 South Hicks Road, Palatine, Ill. 60067 (708) 934-6000
Bench vises, C-clamps, quick-action clamps.

•Woodcraft Supply, Wood County Industrial Park, P. O. Box 1686, Parkersburg, W. Va. 26102-1686 (800) 225-1153
Pivoting-jaw spring clamp.

Drawings above: Charles Miller

Finish Nailers

Pneumatics are a breeze for nailing trim

by Craig Savage

I used to have a jaundiced view of nail guns, especially those used for finish work, because they conjured up images of work done "quick and dirty." My first experiences with air-powered tools were mostly bad ones. The bulky nailers were in for repairs more than they were in the field. The choice of nails was limited, and when the paper strips of nails fell apart, they just became expensive sinkers. More than once I rushed fellow workers to the hospital with accidentally self-inflicted nail wounds.

That was 16 years ago. Today, I wouldn't consider trim jobs or cabinet construction without a finish nailer. The new models are compact, lightweight and well-balanced. By some accounts they are nearly maintenance free, and by any account, they're much improved over the old ones. Attached by its rubber umbilical cord to a portable air compressor, the finish nailer has replaced the hammer on many production finish jobs.

When running most interior trim, you need nails at least 1½-in. long to penetrate the trim, sheetrock and framing member. So for this article, I considered only air-powered nailers that drive a range of small-headed nails at least 1½-in. long. This distinguishes finish nailers from pinners, brad nailers, staplers and framing nailers.

How they work—All air-powered finish nailers work on the same principle. The handle contains a reservoir of compressed air, supplied through a rubber hose from a compressor (*always* use a compressor, *never* use bottled gas to power a nailer because it can cause an explosion). The compressed air drives a piston attached to the driver blade, which punches the fastener into the work. The cylinder holding the piston is capped by a head valve that closes the top of the cylinder. When the nailer's trigger is pulled, the valve opens and air from the reservoir rushes into the cylinder, forcing the chain reaction of piston to driver blade to fastener (drawing below).

The chambers inside nailers are sealed from one another with O-rings, which occasionally wear out and need replacing. Some of the newer nailers have gone to plastic chamber walls. This means that rubber O-rings are rubbing against plastic instead of metal. There is debate in the industry as to whether this is done to increase the life of the tool, to lighten the tool, or simply to cut cost. The rolling diaphragm is another version of the O-ring. It allows for a shorter tool with a lower profile.

In order to meet standards set by the American National Standards Institute

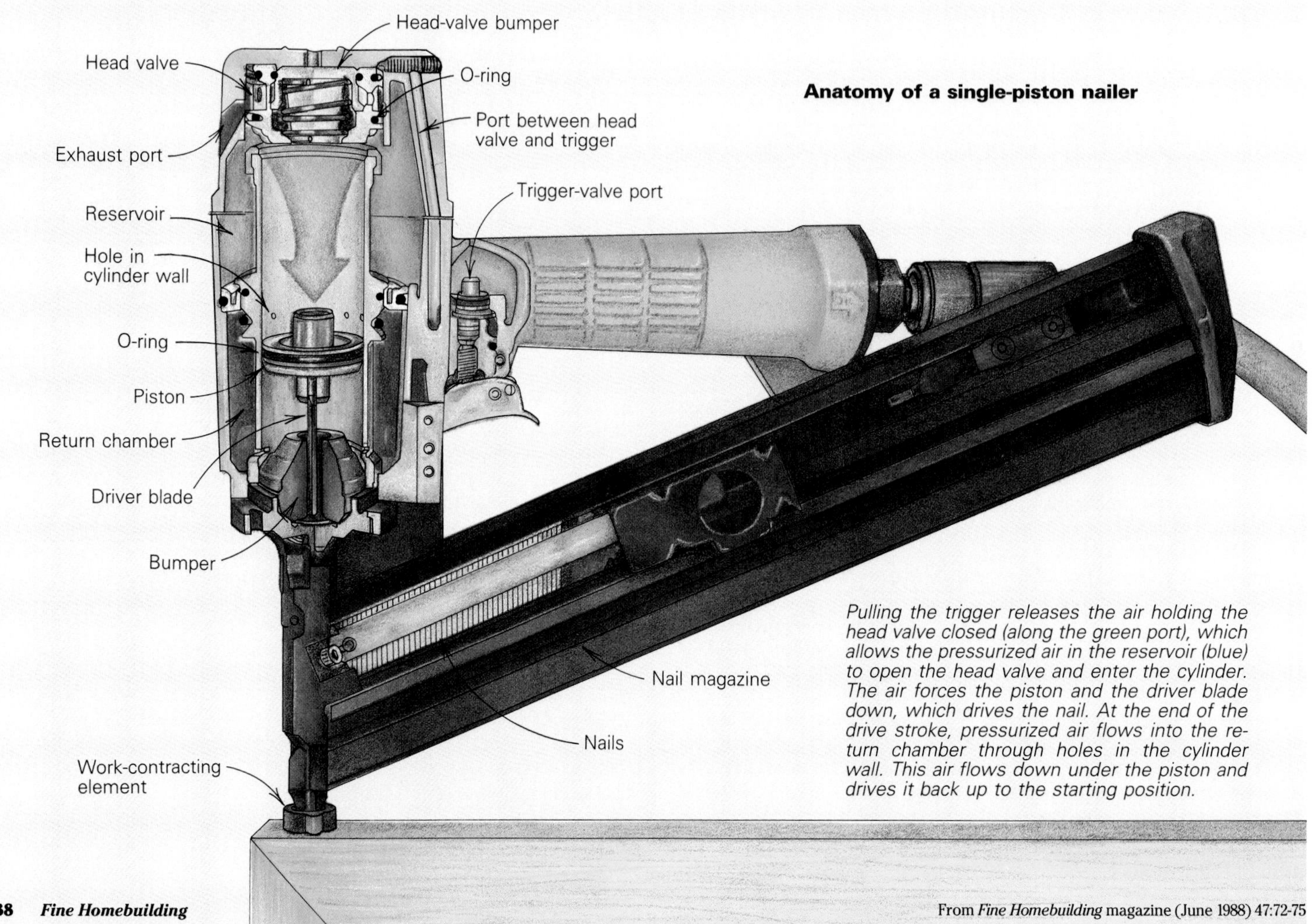

Pulling the trigger releases the air holding the head valve closed (along the green port), which allows the pressurized air in the reservoir (blue) to open the head valve and enter the cylinder. The air forces the piston and the driver blade down, which drives the nail. At the end of the drive stroke, pressurized air flows into the return chamber through holes in the cylinder wall. This air flows down under the piston and drives it back up to the starting position.

From *Fine Homebuilding* magazine (June 1988) 47:72-75

(ANSI), all air-powered tools must have a two-step firing seqence as a safety feature. In addition to the trigger, nailers have a work-contacting element. This is a lever at the nose of the tool that must be depressed against the work at the same time that the trigger is pulled in order to drive a nail. This action prevents the nailer from working like a gun: you can't just point and shoot. The trigger and the work-contacting element have to be used in conjunction; neither can release a fastener by itself.

There are two firing sequences available on most nailers. Contact trip—also called safety-touch trip, dual-action safety or bottom-fire/trigger-fire—is the most common. With this sequence, you can pull and hold the trigger, then bounce the tool along on its nose, firing a nail each time the work-contact element is depressed. This drive sequence is preferred for rapid-fire nail driving.

The other firing seqeuence is called sequential-trip, or trip-then-trigger. After each nail is driven, the trigger must be fully released before driving another one. Also, the work-contacting element must be depressed against the work *before* the trigger is pulled. The sequential-trip sequence takes some getting used to, but it is better suited to trim work than the contact-trip sequence. It allows for exact nail placement, without the risk of a rebounding tool driving a second nail right next to the first. And even more important, sequential-trip models are safer because they will not fire accidentally if the tool comes in contact with something (or someone) while the operator is holding the tool with the trigger pulled (for more on the safe use of air-powered tools, see "Job-Site Safety, *FHB #34*, pp. 51-55). The firing sequence is controlled by the trigger mechanism so you can buy two kinds of triggers and change them yourself, depending on which firing sequence you want.

Nails and magazines—Nails are stored in and fed from spring-loaded magazines, which come in two basic forms—stick and coiled. The stick version holds straight clips of nails, and the coiled version handles rolled belts of nails. No finish nailers on the market today use a coil. Stick magazines can be angled or straight. The angled version is slightly more maneuverable and is able to reach into tighter corners. Some magazines are open, leaving the nails exposed to dirt and grime, others are closed.

Because the nail is contained in the nose of the tool while it's being driven, air-driven nails can be a lighter-gauge wire than hammer-driven nails and still not bend. A standard 8d finish nail is 12-ga. wire, versus 16-ga. for the equivalent length of air-driven nail. The smaller gauge is an advantage for carpenters because the nail is less liable to split the wood. Most nailer manufacturers offer nails in a range of shank styles for their tools, including ring shank, smooth shank and screw shank. Galvanized nails are also available, but I wouldn't use galvanized nails that are made by cutting and forming pre-galvanized wire because the resulting nail is not galvanized on its tip or head and will rust from the inside out. Bostich (see the chart on page 90 for other manufacturers) offers 15 ga. aluminum nails for exterior use. They're more expensive than galvanized nails, but will never corrode since the nail is solid rather than coated. Aluminum will react chemically with some materials, though, so check compatibility before using aluminum nails.

Several carpenters I spoke with mentioned that they were using painted nails (available from most manufacturers) that match moldings. They argue that a painted nail head, set flush, is less obtrusive than a puttied hole.

The process of assembling nails into strips is called collating, and nails are collated in two ways: either straight or at an angle. The straight-collation process is fast and cheap. Nails can be made from square or round wire, and the heads can be T-or L-shaped. These nails are best driven with their heads parallel to the grain of the wood, otherwise they can tear and lift the grain.

The angled collation process is more work, and hence, more expensive (some manufacturers say as much as 35% more expensive). The resulting nail, however, is more like a true finish nail. When it is set into the wood, the rim around the hole is noticeably smaller and neater than when a T-shaped nail head is set. To a finicky carpenter or painter, that kind of detail makes a difference.

I ran across one company, Kotoko of Japan, that makes plastic nails for use with

This compact finish nailer (Spotnail) fits into tight spaces to do its job, leaving the hammer to collect dust. The smaller-gauge nail it drives is less liable than a hammer-driven nail to split the short piece of baseboard pictured above.

Drawing: Michael Mandarano

Finish Nailers								
Manufacturers	**Model**	**Weight**	**Height**	**Length**	**PSI**	**Nail gauge**	**Nail length**	**Nail capacity**
BeA	T-54 155FN	4.5 lb.	9 13/16"	10 3/8"	90-120	13½	1"-2½"	86
Bostich	N-50-FN * N-60-FN *	4 lb. oz. 4 lb. 6 oz.	10" 10 7/8"	13¼ 14⅛"	70-100 70-100	16 15	1"-2" 1¼"-2½"	100 100
Duo Fast	LFN-748 LFN-764 * HFN-880 *	4 lb. 12 oz. 4 lb. 12 oz. 7 lb. 8 oz.	9" 9" 11 7/16"	11 53/64 11 53/64" 12 11/16"	70-120 70-120 50-120	16 16 15	½"-1½" 1"-2" 1½"-2½"	120 120 100
Hilti	BN196B FN200A	4 lb. 3 oz. 4 lb. 14 oz.	8⅞" 9⅞"	9⅞" 12 13/16"	70-120 60-120	16 14	¾"-1 9/16" 1¼"-2"	90 80
Hitachi	NT65A *	4.6 lb.	10"	15"	70-120	16	1¾"-2½"	150
International Staple & Machine	Minor Tip 40 Minor Tip 50	4 lb. 3 lb. 15 oz.	10" 10"	10" 10"	55-70 70-100	16 16	1"-1 9/16" 1"-2"	90 80
Pasload	MU212F *	3 lb. 14 oz.	10¼"	14¼"	80-120	16	1½"-2½"	150
Senco	SFN1 * SFN2 *	4 lb. 2 oz. 7 lb. 8 oz.	9⅝" 11 1/16"	12" 12 13/16"	70-120 80-100	15 14-15	1"-2" 1½"-2½"	104 100
Spotnail	HLB1516 * MNS 5/SP *	4 lb. 9 oz. 8 lb. 8 oz.	10¼" 12 3/16"	11¼" 17"	65-100 85-105	15 10-15	1¼"-2" 2"-3½"	100 50
* Models tested by author								

their Kowa nailer (available from Marukyo U.S.A., Inc., 511 East Fourth St., Los Angeles, Calif. 90013). Nails 1 in. long are the longest available so they're not really finish nailers like others in this article. But they come in a wide range of colors, and these plastic nails will not rust or corrode, can be sanded or cut and will accept stain or paint.

Choosing a nailer—From the trim carpenter's point of view, finish nailers should have special characteristics. They should drive the nail flush or be able to set it to a pre-determined depth. The design of the tool should allow a good view of the nail entry point, so that placement of the fastener is precise. The nose of the tool should fit into tight corners (photo, previous page) and into the tongue on T&G siding, and it should not mar the wood on the rebound. Since they're often used overhead or in other awkward locations, finish nailers should be lightweight and well-balanced.

The ideal finish nailer would drive nails from 1 in. long for cabinets and slender trim to 2½-in. long for hanging doors. Unfortunately, no single tool covers that range of nail lengths. Most companies make at least two sizes of finish nailers (see chart above), dividing the nail capacities around 1½ in., and most carpenters end up buying two nailers.

Some manufacturers claim their tools do not jam, then add that if they do, clearing the jam is quick and easy. After using one, you realize that a compromise has to be made between making the tool easy to open and keeping the thing tightly closed when it should be.

Manufacturers also claim their tools won't mar the wood, but the solutions for preventing this are not always elegant. Pieces of rubber, used to cushion the rebounding tool, can obscure the view of the nail entry point and prevent the nose from fitting into tight corners. I know one carpenter who applies Shoe Goo, or Liqui-sole, to the work-contacting element of his nailers. This stuff is a form of rubber cement used to patch holes in shoes. Twenty-four hours later he carefully cuts the excess away and claims to get about three weeks of cushioned, mar-free operation.

With proper maintenance, which simply means cleaning and oiling, finish nailers should drive ½-million fasteners without major part replacement, according to the manufacturers. I have heard about tools that have fallen off roofs or out of pickup trucks and still not needed major repair for 5 years.

I've talked to a lot of builders who use nailers, but found no clear preference among them for a particular brand. Most were satisfied with whatever brand they were using and would recommend it to others. Service and availability should be the primary factors in your decision of which to buy. In general, finish nailers cost less than framing and roofing nailers, because they're smaller and simpler. List prices range from $300 to $600, depending on make and model. But most dealers will discount the list price, some 20% or more.

Brands and models—In preparation for this article, I invited all the manufacturers of finish nailers I could find to send me samples, though not all responded. I looked over the ones I did receive and drove nails with them. I also lent the nailers to colleagues to try. I learned that driving a nail into 2-in. oak is the finish carpenter's benchmark test of a nailer, so I tried this with all the nailers. Beyond that, I considered qualities like balance, feel and maneuverability. Admittedly, I ended up with subjective observations, some nothing more than first impressions, but they may prove useful if you're sorting through pages of advertising and trying to decide among the host of nailers on the market.

BeA (BeA Fasteners, Inc., 50 Williams Parkway, East Hanover, N. J. 07936) is a West German tool manufacturer that has been making nailers for 25 years. They supply many nailers to the furniture industry. I was never able to get a sample nailer from them for testing.

Bostich (Stanley-Bostich, Briggs Drive, East Greenwich, R. I. 02818) has introduced two new, powerful tools that I liked very much—the Angled-Stick N-50-FN and the N-60-FN. The N-60-FN drives a 15-ga. nail, 1¼-in. to 2½-in. long, which makes it the most versatile of all the nailers in terms of nail length. It has an angled magazine and unique angled drive engine, for more maneuverability. I do wonder if the angle would make it difficult for left-handers to use. It has a dial-a-depth setting that works easily and well, allowing you to leave a nail sticking out a bit, drive it flush, or set it below the surface of the wood. The Angled-Stick features a quick-open nose that pops open automatically when the gun jams. It loads from the rear and is very light—too light, according to some of the carpenters who tried it and complained of its tendency to recoil. The only gripe I have is that the coil spring that pulls the nails toward the nose is very small and is prone to sticking. It reminds me of a ½-in. Stanley tape measure.

Duo-Fast (Duo-Fast Corp., 3702 North River Road, Franklin Park, Ill. 60131) nailers sport a magazine that completely encloses the nails, keeping them free of dirt. The sliding magazine also facilitates removal of jammed nails from the rear of the nose and lets you switch nail sizes quickly, which can be a problem with some of the slide-in-the-back varieties. The work-contact element is coated with clear plastic that protects the work surface and still allows a good view of the work piece. This tool was highly rated by all who used it.

Hilti (Hilti Fastening Systems, P.O. Box 45400, 4115 S. 100 E. Ave., Tulsa, Okla. 74145) is well-known for its powder-actuated tools and on-site service trucks. They sell nailers designed and made in conjuction with Atro, an Italian company. As I was working on this article, Hilti was preparing a new line of finish nailers, but none was available for testing.

Hitachi (distributed by Sivaco Fastening Systems, 615 Falmouth St., Warrenton, Va. 22186) introduced the first pneumatic nailers from Japan. Like many tools today, the Hitachi nailers are designed around modular components. The one I tested (NT65A) is one of the lightest of the finish nailers, which did make me wonder if it would stand up to job-site abuse. Still, it had a nice feel, fit my hand well and seemed quite powerful, easily handling the 2-in. oak test. To remove the nose piece to unjamb a nail, you have only to remove one Allen screw, which seems a good compromise between easy disassembly and accidental disassembly.

International Staple & Machine (P.O. Box 629, Butler, Pa. 16001) sells a tool that is also made by Atro. They turned down my request for a test model.

Pasload (ITW Pasload, Two Mariott Dr., Lincolnshire, Ill. 60015) calls their finish nailer the Mustang Mu212F, and it's well-known among trim carpenters because it works well and has been around for a long time. Pasload reps readily admit that the tool won't pass the 2-in. oak test, which it didn't. But it will nail within ½ in. of corners and is rugged and lightweight. Pasload is working feverishly on downsizing its Impulse hoseless nailers (see *FHB* #33, p. 82), and hopes to have a finish-nailing version soon.

Senco (Senco Products, Inc., 8485 Broadwell Rd., Cincinnati, Oh. 54244) pioneered the first successful pneumatic fasteners for construction back in 1948, and today they are the largest manufacturer of air-powered nailers. Without a doubt, I ran into more Senco users than any other.

Spotnail's MNS 5/SP is the only nailer that drives 16d finish nails, which are available galvanized. So it's a good choice for nailing down a new deck (or renailing an old one, above).

For years Senco's famous SN1 set the standard in the field. Now, the SFN1 and SFN2 have taken its place. These are completely redesigned tools, built with an eye toward lighter weight and longer service. Senco has even refined the SFN1 so that it now drives nails up to 2 in. long (a replacement-magazine kit is available to convert existing SFN1 tools to accept 2-in. finish nails).

The SFN models use a rolling diaphragm instead of O-rings in the main piston seal. The trigger has O-rings but they ride against plastic instead of metal and require little maintenance. Major components are replaced in modules. The module approach to repair means that you don't have to be an expert to fix the tool yourself, but it also means replacement parts are more expensive.

Spotnail (Spotnail, Inc., 1100 Hicks Road, Rolling Meadows, Ill. 60008) has a large selection of finish nailers. The Model HLB1516 I tried was satisfactory in every respect. It passed the 2-in. oak test without its optional power cap, which is supposed to give it even more power. The power cap enlarges the chamber above the piston, resulting in more air volume and more pressure on the piston. Thus, more power is delivered to the driver. Spotnail also manufactures the only nailer that will handle 16d finish nails, the MNS 5/SP. I used it to renail some 2x6 redwood decking and loved it (photo above). Whether by design or by accident, the magazine on this nailer is 16 in. long, making it useful for framing layout.

Safety—Almost 100,000 eye injuries occur in U. S. industries each year. One out of every two construction workers will suffer a serious eye injury during the course of his or her career. By no means do all these accidents result from the use of pneumatic tools. However, the most important precaution you can take is to wear safety goggles at all times when using a pneumatic nailer. In addition, read the owner's manual and understand the tool before using it. Disconnect the air supply before unjamming a nail or doing any maintenance. Do not use more air pressure than the tool is rated for. And as I noted earlier, never use anything but compressor air to power the tool. Speaking from personal experience, the best safety tip I can offer is to carry the tool with your finger off the trigger. □

Craig Savage is a builder and writer living in Palm Desert, Calif. Photos are by the author.

Running Baseboard Efficiently

Simple steps help you make the most of time and materials

by Greg Smith

Let's face it. There is little or no glory in the installation of baseboard. If you want, for instance, to talk about hanging doors, you can probably find plenty of guys who are happy to grant you their expert opinions on the best tools and the most elaborate techniques. But when it comes to installing baseboard, we're back to grabbing a scrap of lumber or an unspent napkin from lunch to record measurements.

The job may go something like this: enter room, plop saw on floor, measure, cut, nail; measure, cut, nail; measure, cut, nail. And you wonder if you'll ever get to the last piece, because it seems like there is always another little piece in some nook or cranny or some space that was missed. It is a job that brings screaming protest from the knees and a hacking voice of discontent from the lungs of the person who fires the nailer that kicks up the dust from the floor adjacent to the workpiece. That may be why, when a team of carpenters is finishing a house, running baseboard is often relegated to the least-experienced person of the group or the low man on the totem pole.

The best way to deal with an unpleasant, though necessary, task is to get it done as quickly as possible. I have seen many carpenters approach the installation of baseboard in many different ways, but I had never seen a system that works very efficiently. That's why I developed a methodical approach that makes baseboard installation fast and efficient.

1. Plan your strategy—The time to run baseboard is before the painter has hidden the location of studs (assuming that you are dealing with drywall) and after the door casing, the built-ins and cabinets and the hardwood or tile floors are installed. In the areas that will be carpeted, hold the baseboard off the floor with a piece of hardwood flooring or other scrap of ¾-in. material—you won't want your beautiful work hidden by the carpet. Though I like to leave a wake of completed baseboard behind me when I am working, that's not always possible. If the bathroom floors have yet to be tiled, for example, I cut my baseboard for the room and set it aside.

2. Set up the saw—I usually set up my 10-in. power miter saw across extra-tall sawhorses (so that I don't have to bend far to see a close-up of the cut) in the biggest room on the floor level on which I'll be working. I'm not as concerned about how close I am to the area to be worked

At the wall. **To run baseboard efficiently, Greg Smith measures several rooms at a time. He prepares a cut list as he goes and marks lengths and cut angles on a preprinted form.**

From *Fine Homebuilding* magazine (August 1992) 76:51-53

At the saw. **With a pile of stock nearby and his list of baseboard dimensions at the ready, Smith parks himself at the saw and cuts down the list. Each piece gets a number that corresponds to the list; later on, he'll be able to mate each length with the correct wall.**

on as I am about having enough room to extend long lengths of material on either side of the saw. With my system, I don't spend a lot of time walking back and forth between my saw and the work area. If there are no large rooms, or if for some reason I cannot use one, I set up instead where I can extend stock out through a door or a window. When all of the baseboard material is spread out near the saw, I'm ready to start work.

By the way, no matter how clean the subfloor is, when you're shooting in baseboard with a nailer, the dust is going to be flying, and your mouth and nose, being close to the ground, are going to scoop up a lot of it. You might want to use a dust mask for this part of the job. In situations where I can free up my left hand, I put it under the place where air is released from the nailer. This keeps the dust from being kicked into the air.

As for what joints to cut, it's up to you whether you miter or cope. The system I use to organize the process won't change the way you work. For purposes of explanation, however, I'll use the example of mitered baseboard.

3. Install the long boards—Tackle the longest walls first—the ones that are longer than the stock you are using. These walls will require the baseboard to be spliced somewhere along its length (drawing facing page, bottom). Start by cutting lengths of stock with a 45° inside miter on the left-hand side and an inside 22½° miter on the right-hand side (assuming you're working from left to right). The splice will occur at the second cut. Later on, you'll be able to take one measurement from the 22½° end to the corner of the wall to complete the wall. Go ahead and install these lengths of baseboard.

4. Take closing measurements—Now take all remaining measurements for three or four rooms at a time. Starting at the door, measure each and every length in the room. Work your way around the room in a clockwise or counterclockwise direction, whichever you prefer. The direction you choose is not as important as being consistent. Use a mechanical pencil (the kind you can get in any grocery or drug store) both for recording your measurements and for marking cutlines later on. The point will always be sharp and consistent, and you won't whittle away time carving up a carpenter's pencil with a razor knife.

Each measurement is recorded on a very simple form that I've designed and carry on a clipboard (photo previous page). The column of blanks is sequentially numbered, and the measurements are recorded in order (drawings facing page). The sequential numbers are important. They will later be recorded on the boards and used to guide you toward correct board placement. You can increase your ability to keep track of what you're doing by drawing a line between each set of measurements when you change rooms. Write the name of the room in the vertical space to the left of the column. In the example, boards #1 to #4 go in the master bedroom, boards #5 and #6 go in the closet.

The solid line to the right of the blanks represents the baseboard as you are looking at it on the wall. On the left and right ends of this line, you will write a symbol to represent the kind of cut required on each end. Use whatever symbols make sense to you—but be consistent. In the small example on the facing page, the straight, vertical line on baseboard #1 means the cut is to be a straight, square cut. This end of the board will butt against the casing around a door. The "2" on the right-hand side of the notation represents an inside 22½° cut. Baseboard #2 starts with a 22½° cut; the slash at the end of the baseboard line indicates a 45° inside cut. Baseboard #4 shows a square cut and a 45° *outside* cut (represented by the O at the end of the line). When working with a stain-grade wood, such as oak, most carpenters like to cope one end to get a tighter fit. I use a C to indicate the end to be coped. You won't find many different cuts in baseboarding, so it won't be hard to memorize the symbols you'll need. If you work with others, you might want to define your symbols on the form so that everyone will be singing from the same song sheet. As for measuring, I simply hook the tape where it's most convenient.

5. Cut each closing board—After you have compiled measurements from several rooms, take your clipboard to the saw and start cutting baseboard (photo left). Each time you cut a board, mark the backside near an end or in the middle (be consistent so that you will know where to look for it later), using the sequential number on your list. Then cross that length off your list. There is no need to write exact lengths on the board, as many carpenters do. For example, if you cut a 22½° inside cut on the left-hand side and 107⅜ in. to the right, make a 45° inside cut; turn the board over, write the number 3 and cross it off your list. Then set the board aside and begin work on board #4.

You need not cut in any particular order. One additional advantage of this method is that you can make very efficient use of your material by taking a little extra time here to avoid waste. Start with your longest pieces, then see what you can get out of the offcuts. You'll have a long list from which to choose.

6. Distribute the boards—Now that you have cut and numbered all of the pieces on your list, you can distribute the pieces. Because you have numbered them in the order in which you measured the walls, it is easy to pick up any random piece and quickly find the place where it belongs. Let's say that the first board you pick up is

M. BED

1.	65½	|—2
2.	48 5/16	2—/
3.	60 3/8	/—|
4.	92 3/8	|—0

CLO

5.	....	——
6.	....	——

Baseboard cutlist. **The sample at left shows how to use the cutlist (below). Marks at the end of each solid line indicate the cuts to be made on each baseboard.**

Baseboard Cutlist

1.	....	——	1.	....	——	1.	....	——
2.	....	——	2.	....	——	2.	....	——
3.	....	——	3.	....	——	3.	....	——
4.	....	——	4.	....	——	4.	....	——
5.	....	——	5.	....	——	5.	....	——
6.	....	——	6.	....	——	6.	....	——
7.	....	——	7.	....	——	7.	....	——
8.	....	——	8.	....	——	8.	....	——
9.	....	——	9.	....	——	9.	....	——
10.	....	——	10.	....	——	10.	....	——
11.	....	——	11.	....	——	11.	....	——
12.	....	——	12.	....	——	12.	....	——
13.	....	——	13.	....	——	13.	....	——
14.	....	——	14.	....	——	14.	....	——
15.	....	——	15.	....	——	15.	....	——
16.	....	——	16.	....	——	16.	....	——
17.	....	——	17.	....	——	17.	....	——
18.	....	——	18.	....	——	18.	....	——
19.	....	——	19.	....	——	19.	....	——
20.	....	——	20.	....	——	20.	....	——

marked 18, and you see that it is the third board cut in the dining room. You can go to the dining room immediately and set the board alongside the third wall from where you started measuring in that room.

7. Nail 'em up—When all the pieces are lying in front of their respective locations, it is time to crawl around the room and nail them up. You may occasionally find that a board or two will need to be recut or trimmed because of walls being out of square or because you measured wrong. This would happen regardless of your approach to the job and will not affect your other cuts. You can either recut as you come to them or write the adjustment needed on the back of the board; e. g., "-⅛" indicates that you need to cut ⅛ in. off of the length.

This system also works well if you like working with a partner. One person measures and installs, the other does all of the cutting. Work this way with a partner only if you like and respect each other and if each of you has a sense of humor. Inevitably there will be discussions about who can't measure right and who can't cut right when boards occasionally come up short or long.

Nothing fancy, but it works—It's a simple system, really, but it does work. You don't have to use a preprinted form, of course, though it does eliminate the need to write out the sequential numbers and draw lines. Besides, it's a heck of a lot easier to write on and read than some of the things that you see carpenters writing on at construction sites.

To save you a bit of setup time, you can photocopy the sample form at right for your own use. Photocopy enough to keep yourself supplied for a few months and keep a few extras in your truck. The columns will allow you to work through the "measure, cut, nail" process 60 times per page. You can also use the form to record measurements for other types of molding and for closet poles. Baseboarding may still be the lowliest job going, but with a bit of organization, you won't be down there quite as long. □

Greg Smith is a general contractor in West Los Angeles, Calif., who specializes in custom home building and remodeling. Photos by Marilyn Ray.

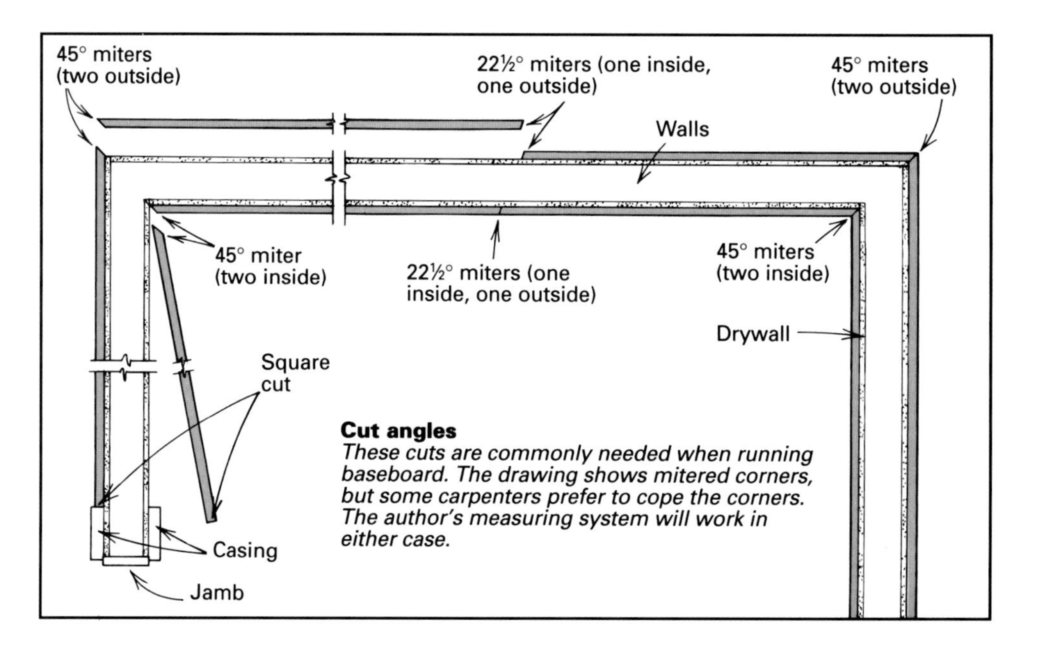

Cut angles
These cuts are commonly needed when running baseboard. The drawing shows mitered corners, but some carpenters prefer to cope the corners. The author's measuring system will work in either case.

Drawings: Bob Goodfellow

Door Hardware

Getting a handle on locksets, latches and dead bolts

by Kevin Ireton

Walter Schlage, a German immigrant working in San Francisco, obtained his first patent in 1909. The idea came to him late one night, as he unlocked his front door and reached in to turn on the lights. The patent was for an electrified door lock that automatically turned on the lights inside a house when the front door was unlocked.

You can't relate the history of the lockset industry without talking about Schlage, Dexter and Yale, men whose companies are still in business. But today, more than 50 other companies now manufacture locksets and dead bolts. From 18th-century reproductions to ergonomically designed lever handles, more door hardware is available today than ever before.

Most door hardware performs at least two basic functions—latching and locking—and many aspire to a third function—looking good. The term lockset refers collectively to the complete latch bolt assembly, trim and handles (knobs or levers). A latch bolt is a spring-loaded mechanism that holds a door closed and may or may not have a lock incorporated into it. A dead bolt, on the other hand, is not spring-loaded and can only be operated with a key or a thumbturn.

Rim locksets and mortise locksets—Originally just wooden bolts that slid across the opening between door and frame, rim locksets were the first broad category of locksets to evolve. They developed into the surface-mounted wrought iron and brass cases common in colonial America.

Eventually, locksmiths realized that doors would look better if the locksets were out of sight. So they mortised the rim lockset into the edge of a door to create the mortise lockset, which prevailed throughout the 19th century. Mortise locksets are still widely used, especially in commercial installations. In residential construction, they're used for exterior doors more ofen than for interior, though models for both uses are available.

Some people consider mortise locksets to be the best type of door hardware available (photo above). Manufacturers can build more

From *Fine Homebuilding* magazine (August 1988) 48:61-65

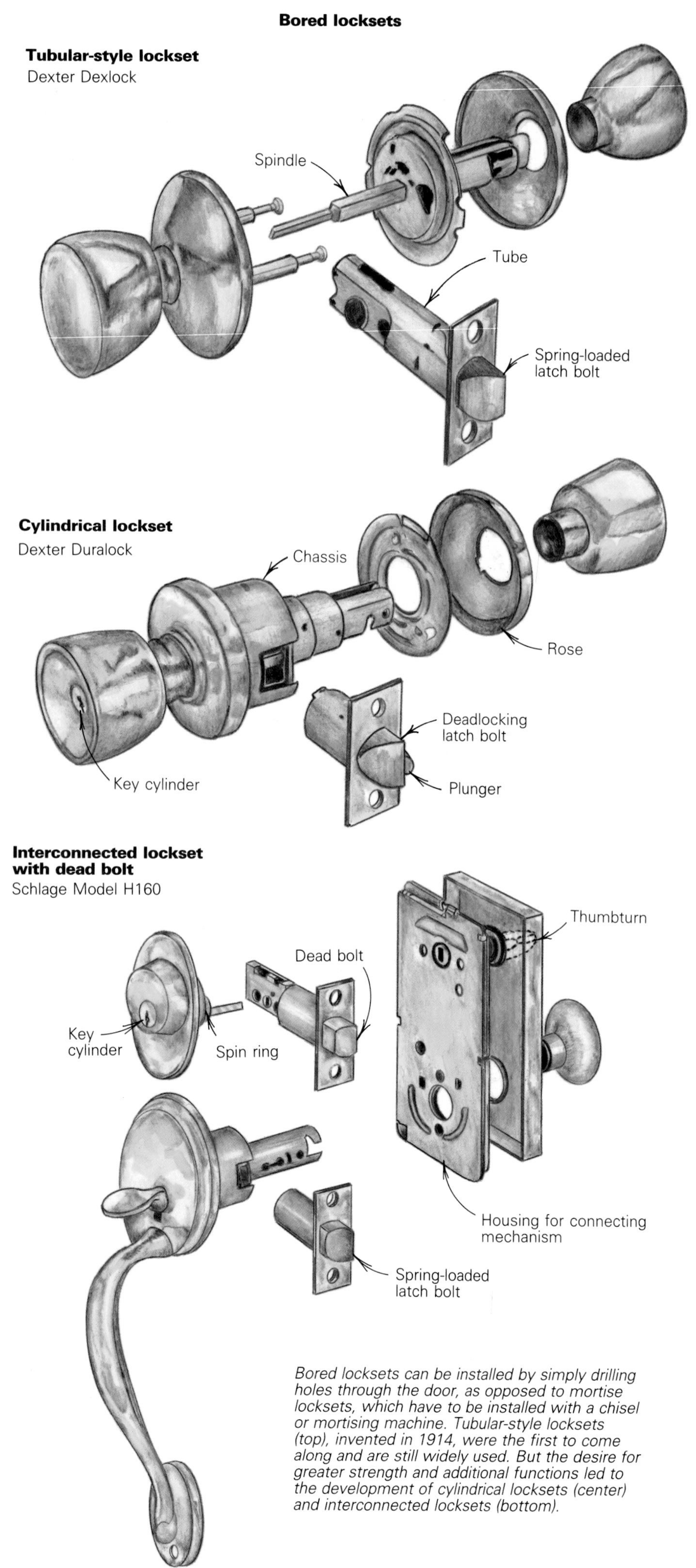

Bored locksets can be installed by simply drilling holes through the door, as opposed to mortise locksets, which have to be installed with a chisel or mortising machine. Tubular-style locksets (top), invented in 1914, were the first to come along and are still widely used. But the desire for greater strength and additional functions led to the development of cylindrical locksets (center) and interconnected locksets (bottom).

strength and longer wear into the larger cases than they can into the bored locksets that I'll discuss later. Also, because the latch and dead bolt are housed together, they can be interconnected to offer a variety of functions. Turning the inside knob on a mortise lockset, for instance, retracts both the latch and the dead bolt—no fumbling for the dead-bolt thumbturn if you're in a hurry to get out. The push buttons, called stop-works buttons, are located below the latch on most mortise locksets and determine whether the latch bolt can be retracted by the outside knob. This type of lockset is also easily adapted for use in extra-thick doors. All that's needed is a longer spindle for the knob.

Aside from the fact that they cost more than other types of door hardware, mortise locksets carry the disadvantage of being tough to install. Cutting a mortise 6-in. long, 4-in. deep and nearly 1-in. wide in a door that's only 1¾-in. thick is tricky and time-consuming. The standard method involves drilling a series of closely spaced holes in the edge of the door and chiseling to clean out between them.

To ease the job, Porter-Cable Corporation (P. O. Box 2468, Jackson, Tenn. 38302-2468) makes a tool called a lock mortiser (model 513). It clamps to the edge of the door and has an integral motor that holds a long mortising bit, much like a router does. Turning a crank on the side of the unit moves the bit up and down along the door and advances the depth automatically. Once it's set up, the lock mortiser works fast, but at a suggested list price of $975, you'll have to mortise a lot of doors to pay it off.

Bored locksets—In 1914, a Michigan hardware manufacturer named Lucien Dexter realized that installing mortise locksets on every door in a house amounted to overkill—nobody needed that much hardware just to keep the pantry door shut, so Dexter invented the tubular-style lockset.

The tubular-style lockset is basically a spring-loaded latch bolt, housed in a tube, that mounts in the edge of a door (top drawing, left). It's operated by a spindle that runs through the end of the tube. A smaller, simpler mechanism than the mortise lockset, the tubular-style lockset is also much easier to install. All you have to do is drill two intersecting holes in the door, one through the face—the crossbore—and the other through the edge—the edgebore (for more on installing bored locks, see p. 28).

The drawback to the tubular-style lockset is that most of the working parts are contained in the narrow tube, so there isn't room to incorporate many functions or a very secure lock. Walter Schlage overcame this problem with his next invention in about 1924, when he developed the cylindrical lockset. Like the tubular-style lockset, the cylindrical lockset incorporates a spring-loaded latch bolt mounted in the edge of the door. But instead of just a spindle passing through the tube, a bigger crossbore hole is drilled (usually 2⅛-in. dia.), and a large chassis is

Drawings: Michael Mandarano

inserted through the door. The cylindrical lockset and the tubular-style lockset are classified together as bored locksets because both are installed by boring holes.

Before the invention of the cylindrical lockset, latching and locking were performed by separate mechanisms. But the large chassis on his new invention gave Schlage enough room to install the key cylinder right in the knob, so that locking and latching could be combined in one mechanism (for more on key cylinders, see sidebar this page). Schlage also developed the button-lock for the inside doorknob, an idea he got from push-button light switches in use at the time.

Among later refinements to the cylindrical lockset was the deadlocking latch bolt. The familiar trick of retracting a latch bolt by slipping a credit card between the door and jamb is known in the trade as "loiding a lock," *loid* being short for celluloid. The deadlocking latch bolt thwarts this practice. With the door closed, a spring-loaded plunger, located alongside the latch bolt, is held in by the strike plate, and the latch bolt can't be retracted. This requires a reasonably conscientious installation. Otherwise, the plunger can slip into the hole for the latch bolt, which defeats the purpose.

Dead bolts—"Don't rely on a lockset for security." I've heard that from a number of people in the industry. Projecting out from the door like it does, a key cylinder in a doorknob is vulnerable to being smashed or sheared off. If you want security, you have to have a dead bolt. Some insurance companies even offer you a 2% or 3% discount on homeowner's insurance if you have dead bolts on all your exterior doors. Still, I was reminded by a locksmith that, "One-hundred percent security is not available on this planet. Anything that's manmade can be defeated. By using dead bolts, you're simply trying to make access more difficult."

Dead bolts should have a full 1-in. throw, which means the bolt will extend 1 in. into the door jamb. At one time, shorter throws were common, and some companies may still make them. But if the dead-bolt throw is less than 1 in., it's easy for an intruder to prize the door jamb away from the door and free the bolt.

Many companies run steel inserts through the center of their dead bolts so that if someone tries to hacksaw them, the inserts will roll under the blade instead of allowing it purchase to cut. It's a great idea in theory, but no one I asked had ever heard of a burglar hacksawing a dead bolt; there are too many easier ways to break into a house. Peace of mind, I learned, is a major factor in selling home security.

Most dead bolts include a key cylinder outside and a thumbturn inside the door. But dead bolts are also available with a double cylinder, which means you need a key to unlock it from either side of the door. Double-cylinder dead bolts are used in doors that have windows so an intruder can't break the window, reach in and unlock the door. In emergencies, though, they can pose a safety hazard, and their use is restricted by code in some areas.

The weak spot in a dead-bolt installation is the strike—the metal plate mounted in the door jamb that the dead bolt goes into. It should be mounted with 2-in. or longer screws that penetrate the trimmer stud behind the jamb. Some companies supply long screws with their dead bolts, but some don't, and often you have to angle the screws to sink them solidly into the framing. But now a few manufacturers have begun to offer dead-bolt strike plates with holes that are offset toward the center of the door jamb.

Many dead bolts come with a metal or plastic housing, called a strike box or dust box, that installs behind the strike and keeps the dead bolt contained inside the jamb. Installing it requires extra work with a chisel, and I'll admit to throwing away more strike boxes than I've installed. I thought their only purpose was to make the hole in the jamb look neat. But it turns out that strike boxes serve a couple of important purposes. For one thing, they keep debris from getting into the strike hole and jamming the bolt. But

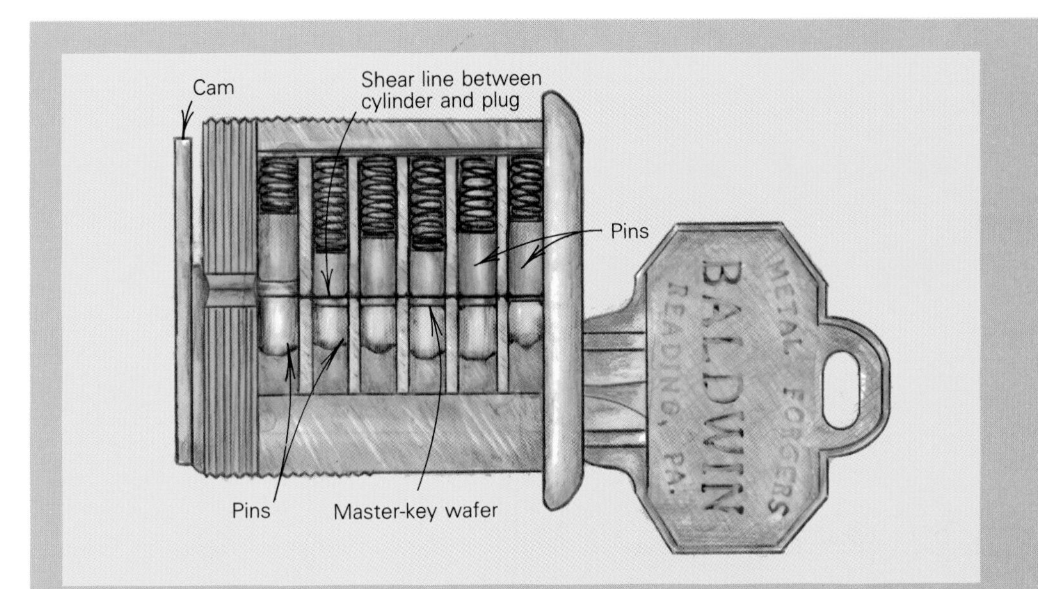

Pin-tumbler cylinders

The oldest known lock, found at the ruins of the palace of Khorsabad in present-day Iraq, is over 4,000 years old and operates on the same principle as most door locks today—the pin-tumbler. This lock and others like it were operated by a wooden key shaped something like a large toothbrush, but with wooden pins in one end instead of bristles. When the key was inserted and lifted up, the pins on the key aligned a set of pins inside the lock that allowed a wooden bolt to be retracted. This type of lock was common in Egypt and is called an Egyptian lock. Subsequent civilizations developed new locks based on different principles, and pin-tumblers weren't used again until 1848 when Linus Yale "reinvented" them.

Yale developed and patented a simple key-operated lock based on the pin-tumbler principle. His son later refined the lock by encasing it in a small metal cylinder with a rotating internal plug. Although plenty of locks have been invented since—combination locks, push-button locks, credit-card locks and even electronic locks—the pin-tumbler cylinder is still the heart of most door locks used today.

***How it works*—A series of holes, usually five or six for a door lock, are drilled through the top of the cylinder into the plug. At least two pins of different lengths, plus one spring, are inserted into each hole. Some cylinders have brass pins, but the best cylinders use nickel silver, which is a harder metal than brass and assures that the key will wear out before the pins. By varying the combinations of pin lengths, a manufacturer can key any given cylinder literally millions of different ways.**

The line between the plug and the cylinder is called the shear line. Any misalignment of the pins across the shear line prevents the plug from rotating in the cylinder, and hence, the lock from unlocking. But when the right key is inserted in the keyway, the pins slip into the gullets of the key's serrated edge, which raise them to just the right height, aligning their ends along the shear line (drawing above). The plug is then free to turn in the cylinder, and when it does, a cam attached to the back of the plug throws the bolt.

A master key is one that will work two or more cylinders that are keyed differently. Cylinders accommodate a master key by means of master-key wafers inserted with the regular pins. These wafers create additional points where joints between the pins align at the shear line and the plug is free to move.

Most key cylinders are assembled by women because the pins are very tiny, and women, with their smaller hands, can assemble cylinders much faster than men can. *—K. I.*

their main function is to serve as a depth gauge. If the dead bolt is less than fully thrown, only ¾ in. for instance, it will easily spring back into the door when you tap it—not a very secure situation. Installing the strike box ensures that the dead bolt can be fully thrown.

The large beveled housing surrounding the outside of a dead-bolt cylinder is called a spin ring. It protects the key cylinder from being removed with a pipe wrench or locking pliers. When buying a dead bolt, be sure to get one that has a solid metal insert for the spin ring. Hollow rings can be crushed easily.

Weiser Lock (5555 McFadden Ave., Huntington Beach, Calif. 92649), makes a combination lockset and dead bolt called the Weiserbolt. It works like a standard lockset, until you turn the key in the lock or twist the thumbturn. Then the latch bolt extends a full 1 in. into the jamb. But while the Weiserbolt is more secure than a standard lockset, it isn't as secure as a separate dead bolt because the key cylinder in the projecting knob is still vulnerable. It does, however, provide additional security without requiring an extra hole in your door.

Interconnected locksets—In order to combine one of the functions of a mortise lockset with the easy installation of a bored lockset, the Schlage Lock Co. introduced the interconnected lockset in 1967. It's an entrance lockset that links the latch and dead bolt so that both will retract by simply turning the inside handle. The mechanism connecting the dead bolt and latch bolt is contained in a thin rectangular housing that mounts on the inside face of the door (bottom drawing, p. 96). Interconnected locksets cost more than separate locksets with dead bolts, but they are more convenient to use.

Interior locksets—Locksets for interior doors don't usually have a keyed lock, but may instead have a simple button lock. These are called privacy locksets and are used for bedrooms and bathrooms. Privacy locksets have an emergency release so that a locked door can be opened from the outside either with a special tool that comes with the lockset or with a small screwdriver.

A passage lockset is one with no locking function at all; turning the knobs or lever handles on either side of the door will always retract the latch. Some manufacturers offer a special closet lockset that has a knob or lever handle on the outside, but a simple thumbturn on the inside. These are usually available only in more expensive lines, because if you're paying $30 or $40 for a solid brass doorknob, you may not want waste it inside a closet. A dummy knob is a handle with no lockset and is often used on the stationary half of a pair of French doors.

Quality and standards—Many companies produce a broad range of locksets that vary, not just in style and finish, but in quality. Two locksets made by the same company may look exactly alike behind the cellophane windows on their packages, but one might cost $25, and the other, $70.

Among the factors that determine the cost of a lockset are whether the key cylinder is made of turned brass (which can be machined to close tolerances and is long-wearing), a zinc die casting, or plastic. The steel parts inside a good grade of lockset are treated for corrosion resistance; cheaper ones aren't. One locksmith I talked to suggested picking up the locksets to gauge their weight. The heavier one, she told me, is probably better. Someone else suggested working the mechanism and trusting my instincts about which feels smoother.

It's likely that the $25 lockset is a tubular-style and the $70 lockset is a cylindrical. Following World War II, manufacturers developed the ability to install key cylinders in tubular-style locksets. But in general, tubular-style locksets don't meet the same standards that cylindrical locksets meet.

The Builder's Hardware Manufacturer's Association, Inc. (BHMA, 60 East 42nd St., New York, N. Y. 10165) sponsors a set of lockset standards published by the American National Standards Institute (ANSI) and certifies locksets according to those standards. Products from companies that subscribe to the standards (not everyone does) are selected at random and without notice and are tested for strength, performance and finish by an independent laboratory. The grading that results is rather involved, but a few examples will give you some idea of what goes on.

Bored locksets (ANSI A156.2) are certified in three grades. Grade 1 means that the lockset is suitable for heavy-duty commercial use. Grade 2 means it's suitable for light-duty commercial use. And grade 3 means it's suitable for residential use.

In one test, a lockset is attached to a door that's opened and closed by machine at a rate of 10 times per minute. A grade 1 lockset

The photo above shows the stages a lever handle goes through in the hot-forging process. Once cut to length, pieces of bar stock, called billets, are heated to 1,400° F. The red-hot billets are placed in a die, slammed by a forging press, and the shape of the lever emerges. Excess brass (flashing) squeezes out around the edges of the forging and is trimmed on a punch press. The lever then goes from grinders to polishers to buffers and finally to the lacquer spray booth.

has to be operational after 600,000 cycles, grade 2 after 400,000 and grade 3 after 200,000. In another test, a 100-lb. weight is swung into a closed door from a given height to see if the latch bolt will bend or break, allowing the door to open.

Should you consider installing grade 2 locksets on a house? Not necessarily. Grade 2 hardware will stand more abuse than grade 3, but the difference might only be whether it takes five minutes to break into the house or three minutes. In either case, the weak link is most likely to be the door and the jamb, not the hardware. Grade 2 locksets will certainly be more durable than grade 3. But whether they will see enough use in a house for the difference to become apparent is another question.

Not all manufacturers participate in the certification program, and not every product of the participating manufacturers meets the minimum standards. To make matters worse, most companies don't advertise their certification on packaging. They do, however, include it in their catalogs, so if you're interested, get a catalog. Or better yet, contact the BHMA and for $2.50, get a copy of their "Directory of Certified Locks and Latches."

Reaching for the brass knob—Brass, an alloy composed of 60% to 70% copper mixed with lead and zinc, is the premier material for door hardware. It offers the right combination of strength, workability and corrosion resistance. According to builders and hardware dealers I talked to, some of the finest brass knobs and handles generally available are made by the Baldwin Hardware Corporation (841 Wyomissing Blvd., Reading, Penn. 19603). If you're converting the room above your garage into a rental unit to help meet your mortgage payments, Baldwin locksets and handles probably aren't for you. Their solid brass hardware is expensive, but considered worth the cost by those who can afford it. I visited Baldwin's plant to find out why.

The heart of Baldwin's operation is hot forging, the same process used by blacksmiths. Raw material, purchased from brass foundries in the form of bar stock, is cut to length, and the pieces, or billets, are conveyed into furnaces and heated to 1,400° F. The red-hot billets are then placed in dies and whomped into shape by a huge forging press. Excess material called flashing seeps out around the forging and is later trimmed on a punch press (photo, facing page).

Baldwin touts hot forging over the sand casting used by other companies by pointing out that the resulting pieces are denser, stronger and smoother, lacking pit holes and air pockets. But in the context of residential use, they aren't necessarily more durable or more secure. In this price range, the difference comes down to aesthetics—how it looks and feels, which is why Baldwin takes such pains with polishing, buffing and finishing.

The biggest challenge faced by Baldwin and by any manufacturer of expensive hardware lies in the nature of custom homebuilding. Door hardware is one of the last things to be installed in a house, and if the project is over budget, hardware is a common target for cost-cutting. To learn what a cost-cutting homeowner or builder might use instead of solid brass, I visited the Dexter Lock Company (300 Webster Road, Auburn, Ala. 36830). Dexter does sell a designer series of forged brass door hardware, but I wanted to learn how they make the ubiquitous hollow doorknob that most of us reach for every day.

At Dexter, doorknobs are formed on a sixteen-stage transfer press. The brass is 8 in. wide and .028 in. thick and is fed to the press from coils as big as wagon wheels. After cutting it into 8-in. squares, the machine transfers the brass from one die to the next, forming the familiar shape a little more each time and trimming off the excess. Toward the end of the process, the knob is filled with fluid and expanded by hydraulic pressure to create the final shape. The whole process takes less than a minute.

The big difference between solid brass and hollow brass doorknobs ends at that point. Both get much the same white-glove treatment as they're passed from polishers to buffers and then conveyed on racks through the lacquer spray booth and drying ovens.

Exotica and where to find it—The Schlage Lock Co. (2401 Bayshore Blvd., San Francisco, Calif. 94134) recently introduced something called a Key N' Keyless Lock. It's a lockset with optional dead bolt that can be opened without a key simply by twisting the knob left and right in sequence. A British company, Modric, Inc. (P.O. Box 146468, Chicago, Ill. 60614) makes lever handles in 355 colors, with cabinet hardware and bath accessories to match (photo below, center).

Valli & Colombo (P.O. Box 245, 1540 Highland Ave., Duarte, Calif. 91010), an Italian company, recently introduced a designer line of door handles developed for the disabled (photo below, left). Meroni, another Italian company, makes a push-button doorknob (distributed by Iseo Locks Inc., 2121 W. 60 St., Hialeah, Fla. 33016) that opens when you squeeze it (photo below, right).

Normbau Inc. (P. O. Box 979, 1040 Westgate Dr., Addison, Il. 60101) and Hewi, Inc. (7 Pearl Ct., Allendale, N. J. 07401) both make colorful nylon-coated door hardware that's tough and won't corrode or tarnish.

You probably won't find these products in local hardware stores or lumber yards. Look in the Yellow Pages under "architectural or builder's hardware," or go to a locksmith. If you are interested in antique or reproduction hardware, check out the listings in the *Old-House Journal Catalog* (Old-House Journal Corporation, 69A Seventh Ave., Brooklyn, N. Y. 11217. $15.95, softcover). You can also write to the Door and Hardware Institute (7711 Old Springhouse Road, McLean, Va. 22102-3474) for a copy of their *Buyer's Guide, 5th Edition*. It will cost you $35, but you'll get the most comprehensive list of door-hardware manufacturers that I've seen. □

Kevin Ireton is an associate editor with Fine Homebuilding.

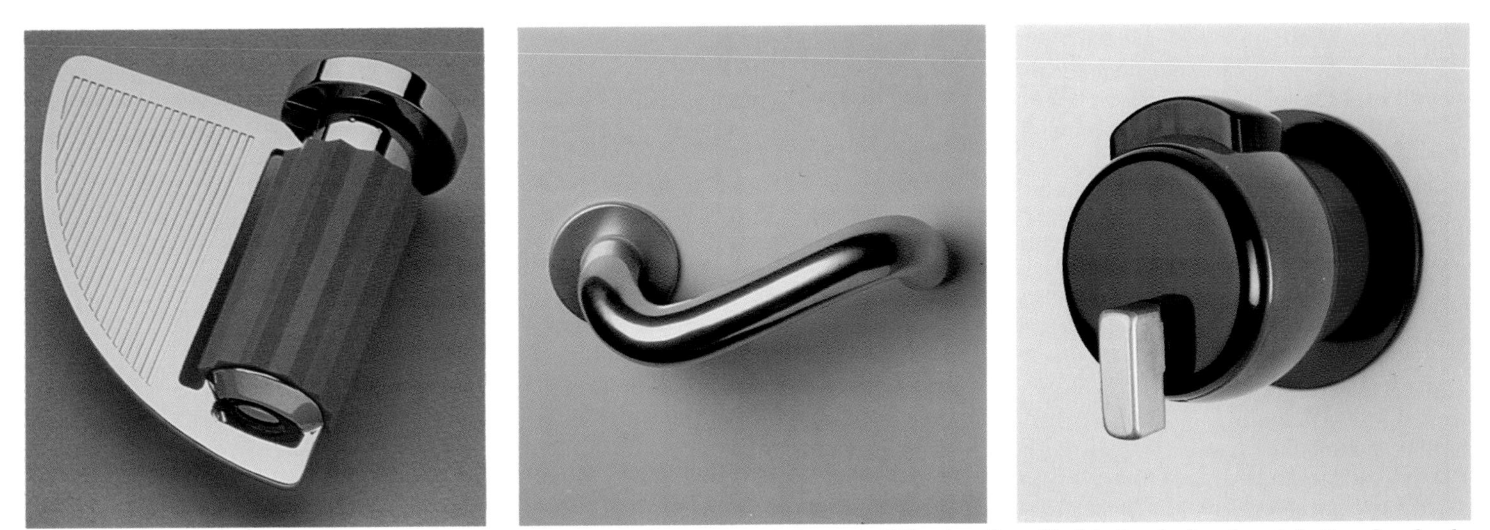

Designer handles for the disabled (left), made by Vali & Colombo, ergonomic levers in 355 colors from Modric (center) and push-button doorknobs from Meroni (right) are just a few of the European door handles recently introduced in the United States.

Building Interior Doors

Using a shaper to produce coped-and-sticked frames and raised panels

by Joseph Beals

The dark, tired hallway in our Cape Cod-style house badly needed a face-lift. Like the rest of the interior, the hall had been built on a tight budget. With its aging acoustical-tile ceiling, cheap plywood paneling, assortment of ill-fitted stock trim moldings and five pre-hung hollow-core doors, there was little worth saving. My helper and I gutted the hall of everything but the existing door jambs and vinyl flooring. To give an illusion of height while creating an eye-catching detail, we installed a vaulted ceiling using drywall and curved 2x6 scabs fastened to the ceiling joists. We also applied ½-in. drywall to the walls.

But the heart of the hallway remodel was the construction and trimming out of five solid-wood, frame-and-panel doors (photo right): two 30-in. wide bedroom doors, a 30-in. wide bathroom door, a 34-in. wide cellar door and an 18-in. wide closet door. Four of the five existing doors had been hung on split jambs, with the adjustable portion of the jambs located on the hall side. This enabled us to reset the jambs flush with the new drywall. The closet door was hung on a conventional jamb, which we removed and reinstalled flush with the drywall.

Picking the pattern—The typical commercial frame-and-panel door is the six-panel Federal-style whose almost-flat panels are raised in the most minimal sense of the word. This generic reproduction style would be out of place in our remodel. Rather than subcontract the manufacture of custom doors to a local shop, I decided to design and build my own.

I drew principally on two references for design information: a turn-of-the-century English work called the *Handbook of Doormaking, Windowmaking and Staircasing* (reprinted by the Sterling Publishing Co., Inc.; 212-532-7160); and "Making Period Doors," an article that appeared in *Fine Woodworking* magazine (*FWW* #71, June/July 1988, The Taunton Press, Inc.). After sketching several options, I designed a four-panel Greek-Revival door with panels raised flush to the frames on both sides. This not only creates a very strong play of light and shadow, but it allows the convenience of a single thickness of stock for all door parts. For the master-bedroom door at the end of the hall, I substituted two round-top, leaded-glass panels for the top two raised panels. The narrow closet door has just two raised panels, one above the other.

Although there would be three different door sizes, the dimensions of stiles, rails and mullions would be consistent throughout (drawing facing page). I made the mullions ½ in. narrower than the stiles, and the middle (or *lock*) rails ½ in. narrower than the bottom rails, to give a balanced appearance. The centers of the lock rails are located 36 in. from the bottoms of the doors to produce classic Greek-Revival proportions.

The construction of the doors would echo that of the typical commercial frame-and-panel interior wood door—that is, cope-and-stick joinery reinforced with dowels where rails meet stiles. Coping and sticking involves the cutting of a continuous decorative molding (the *sticking*) along the inside edges of the stiles, rails and mullions, then coping the ends of the rails and mullions so that the end cuts are a perfect inverse of the sticking and fit snugly against it. This method surrounds each door panel with molded edges that appear to be mitered at the corners.

A century ago, cope-and-stick joints were cut using a variety of molding planes, chisels and gouges. Now, they're typically produced with a shaper. I own a ¾-in. shaper and a set of carbide cope-and-stick cutters with an ovolo molding profile on them—just what I needed for this job.

The choice is poplar—Most commercial paint-grade interior doors are made of pine or fir. I built mine out of 8/4 poplar planed to a thickness of 1⅜ in., a standard dimension for interior doors. Poplar is a relatively stable, straight-grained hardwood that's harder than most softwoods, works easily and takes a painted finish extremely well. Poplar isn't particularly rot-resistant, however; I wouldn't recommend it for exterior doors.

The poplar cost $1.40 per bd. ft. I don't own a thickness planer, so my local supplier planed the poplar for me for an extra 15¢ per bd. ft. To reduce waste and help cut costs, I planned to make no spare parts, a risky practice that leaves no room for error.

Roughing out the pieces—Each door was built to the exact dimensions of its existing opening, then trimmed to fit. I ripped the door parts on a table saw and cut them to length on a radial-arm saw. When determining the lengths of the rails and mullions, I allowed for the depth of the coping profile—in this case 17⁄32 in. The mullions were left long until the stiles and rails were machined and assembled dry. That allowed me to lay out the exact lengths of the mullions before cutting them.

Doweling—Dowels replace traditional mortise-and-tenon joints and are a critical adjunct to cope-and-stick joinery. Doweling typically precedes coping and sticking because layout and drilling are most easily accomplished before the frames are molded. I used 1-in. dia.

A classical face-lift. **Trimmed with custom casings, this shop-built, frame-and-panel door is one of five new doors that highlight the author's hallway remodel.**

Photo: Bruce Greenlaw

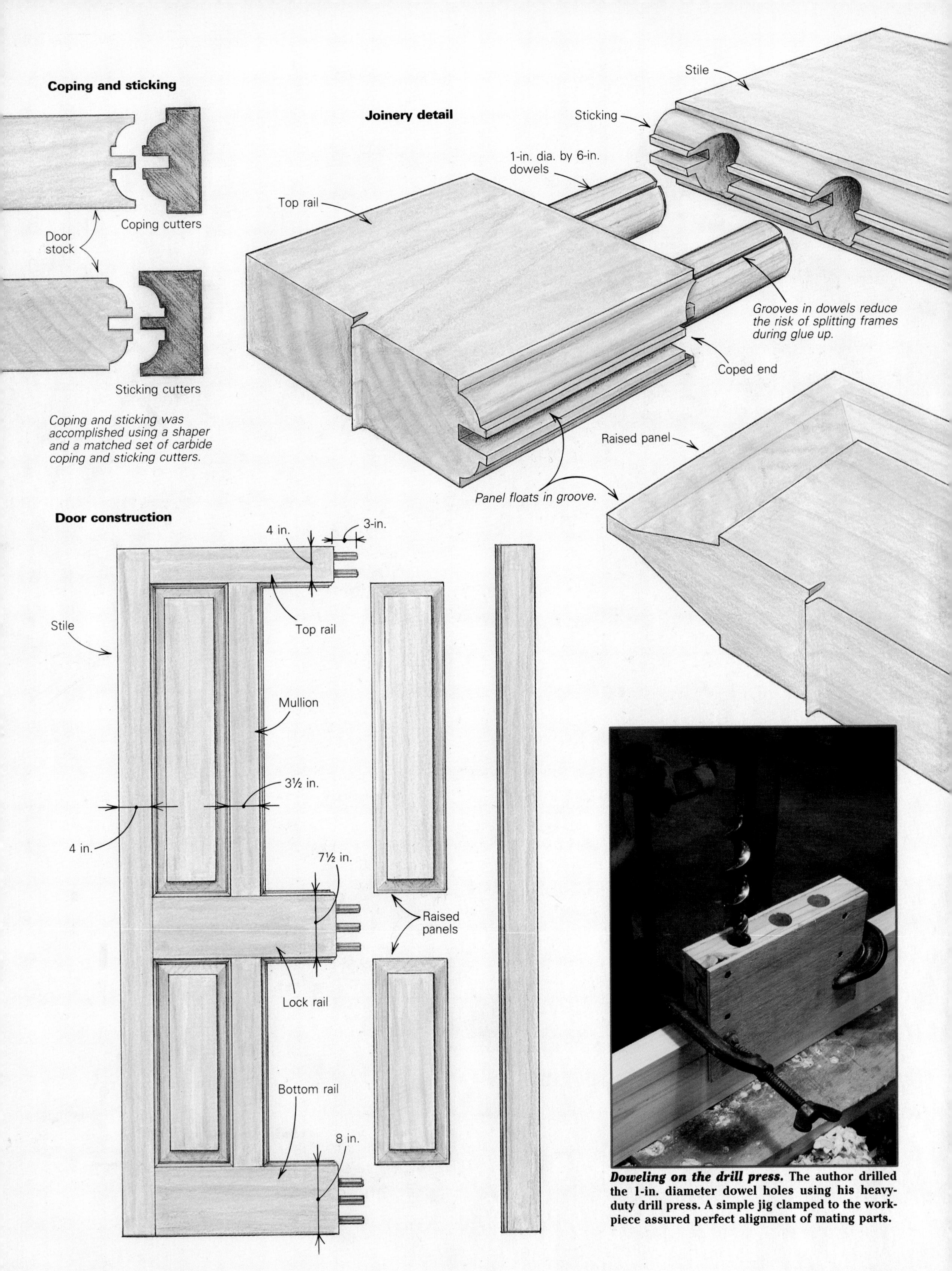

Doweling on the drill press. **The author drilled the 1-in. diameter dowel holes using his heavy-duty drill press. A simple jig clamped to the workpiece assured perfect alignment of mating parts.**

Drawings: Bob Goodfellow

by 6-in. long dowels: two at each end of the top rails and three at each end of the bottom and middle rails. Spaced 2¼-in. o. c., the dowels are held at least ⅝ in. away from the edges of the rails to allow milling and trimming of the doors without cutting into the dowels. The cope-and-stick joinery alone suffices for anchoring the ends of the mullions to the rails.

I drilled 3⅛-in. deep dowel holes using a shop-made guide in tandem with a heavy-duty drill press to ensure precise alignment of the mating parts (photo, p. 101). The dowels were cut from 3-ft. lengths of standard 1-in. dowel stock, but not before I ripped on the table saw a pair of ⅛-in. deep saw kerfs 180° apart. These grooves would allow glue to squeeze out the bottoms of the dowel holes during glue up, relieving pressure and reducing the risk of splitting the frames. When cutting the grooves, I guided the dowel stock by feeding it between the standard rip fence and an auxiliary fence, which were installed about 1 in. apart on opposite sides of the sawblade. Because the depth of cut was a meager ⅛ in., I felt safe using this method for cutting the saw kerfs. I would devise a different method for making deeper cuts in round stock, however, because the stock *could* rotate while being fed through the saw and kick back.

Once I cut the dowels to length, I drove each one through a 1-in. dia. hole that I drilled through a ⅜-in. thick steel plate, ensuring dimensional consistency (the dowels initially measured slightly more than 1 in. in diameter and were slightly oval). Finally, using a stationary disk sander, I tapered the ends of the dowels to ease installation. The completed dowels were then set aside until glue up.

Coping and sticking—There are a variety of carbide cope-and-stick shaper cutters on the market. Most come in matched sets: a sticking set and a coping set. The sticking set typically consists of two molding cutters separated by a straight cutter; the straight cutter mills a groove for holding door panels. The coping set consists of two molding cutters ground to an exact inverse of the sticking cutters and includes a spacer instead of a straight cutter (detail drawing previous page). The spacer leaves a tongue on the ends of the rails and mullions that engages the panel groove produced by the sticking cutters.

My carbide cutters, which are made by Freud (218 Feld Ave., High Point, N. C. 27264; 800-334-4107), come with shims for fine-tuning the fit between the coped tongue and the panel groove, but I've never had to use them.

Before coping and sticking the door parts, I experimented on wood scraps, adjusting the setups where necessary. When cutting the door parts, I cut all the copes first, then the sticks. That way, any splintering produced by the coping cutters at the trailing edge of the stock (common when cutting wood across the grain) would be removed by the sticking cut. I also limited tearout during coping by backing up the work with a wood scrap. When sticking the stiles, I used infeed and outfeed rollers to help support the ends, and hold-downs (steel clips) to keep the work flat against the table.

Panel raising—Panel dimensions can be calculated from plan drawings, but that method invites mistakes. I prefer to knock the door frames together dry, without dowels, and to measure for the dimensions of the panels directly from the frames. I measure to the bottoms of the panel grooves, then subtract about ⅛ in. per foot of panel width to allow for seasonal expansion of the wood (for more on the seasonal movement of solid wood, see *FHB* #69, p. 54).

The panels were raised on the shaper using a carbide raised-panel cutter. Prior to shaping, however, I roughed out the bevel cuts on the table saw to remove the bulk of the material. This allowed me to run the finish cut on the shaper in one easy pass, reducing both machine time and wear on the panel-raising cutter. The panels are raised on both sides, so I adjusted the finish cut to leave the correct tongue thickness (about ¼ in.) where the panel edge is captured in the door frame.

Safety on the shaper—For convenience with the long, heavy door panels, I ran the shaper cutter submerged (rotating clockwise below the stock). This is opposite the usual procedure, in which the cutter rotates counterclockwise *above* the stock. My method has two advantages: the stock shields the cutter from fingers, and if the panel is inadvertently bumped on the shaper table, the only result is a crown that can be removed quickly with a second pass. The only drawback to this method is the mass of chips that can accumulate around the cutter, often in sufficient volume to heat the table and bog down the motor. I stopped the shaper after raising each panel and blew out the waste with compressed air. This interruption can be annoying, but it leaves you with perfect panels and a full complement of fingers.

Gluing up—I assembled the doors with aliphatic-resin glue, which is sufficient for interior work. When assembling the four, four-panel doors, I began by gluing up sub-assemblies of three rails and two muntins. Then I slid the panels into place, brushed glue into the dowel holes, drove the dowels into the stiles and, finally, drove the stiles onto the rails. Using a pair of pipe clamps across each rail, one over and one under, I drew the joints up tight and true. On the two-panel closet door, I simply assembled a stile and the three rails, inserted the raised panels, drove on the opposing stile and clamped the door tight.

Because of the long bearing surface at the ends of the wide rails, the doors drew up square without effort. If they hadn't, I would have corrected them by skewing the clamps off square to pull the doors into alignment and checking for square by measuring the diagonals (diagonals that are equal in length indicate a square door). The long dowels kept the joints relatively flat, and any deviations were straightened out by tweaking the clamping pressure. Shortly after assembly, I moved all the panels within their grooves to ensure that they weren't inadvertently caught by squeezed-out glue inside the joints.

Coping in the round—To add light at the end of the hall, I fitted the master-bedroom door with a pair of round-top, leaded-glass panels that I made in the shop (photos facing page). The semicircular panel heads echo the vaulted ceiling above, and the colored glass casts a soft, almost ecclesiastical light at the dark end of the hall without compromising privacy in the bedroom.

The door was assembled like the others but without raised panels in the upper two openings. After assembly, I glued up four simple filler blocks, coped them on the shaper and glued them into the upper corners of the panel openings. Next, I laid out curved cutlines on the blocks using a pencil compass and rough-cut the curves with a jigsaw. That done, I molded the rough-cut curves on the shaper, guiding the cut with a ball bearing stacked on top of the sticking cutters and a plywood template screwed to the door.

To permit the installation of the leaded-glass panels, I removed the molded edge on the back side of the door using a router chucked with a straight bit and guided by another ¼-in. plywood template clamped to the door. The glass panels are centered in the openings with several daubs of clear silicone caulk and locked in place with removable molded wood stops.

I made the straight stops by sticking the jointed edge of a poplar board, then ripping off the molded stop on the table saw. To make the curved stops, I began by gluing up a block of scrap poplar, bandsawing a radius in it and sticking the radiused edges on the shaper using the same bearing/template system used for sticking the curved edges in the door. The curved stops were then cut from the blocks on the bandsaw and dressed to a precise fit on their convex sides using the disk sander.

With the glue ups completed, it was time to clean up the doors and smooth the joints. Some doormakers accomplish this with a belt sander, but I prefer to use a sharp bench plane.

Making casings—Before hanging the doors, we made and installed poplar casings to match. The casings (photo, p. 100) are designed to echo the classic colonial detailing in the recently renovated living room adjoining the hall. I milled a double-ogee pattern into the 4-in. wide side casings using a single-knife molding head on the table saw. This molding head is an old tool that looks primitive, but it can produce a very fine finish. Because there is only one knife to grind for any particular segment of a profile, custom patterns can be set up with little difficulty. I achieved the desired profile by making two

From *Fine Homebuilding* magazine (February 1992) 72:50-53

1. Filling the corners. **To accommodate the arched, leaded-glass panels in the master-bedroom door, the author glued up four filler blocks, coped them on a shaper and glued the blocks into the panel openings.**

2. Shaping round the bend. **Next, Beals rough-cut the arches using a jigsaw and molded the radiused edges on the shaper, guiding the cut with a ball bearing riding against a plywood template screwed to the door.**

3. Routing recesses. **To allow installation of the panels, the author then removed the sticking on one side of the door using a router guided by a second plywood template screwed to the door.**

4. Curved stops. **Curved stops were made by bandsawing their inside radius in a poplar scrap, molding the radiused edge on the shaper (guided by a template), bandsawing the outside radius and smoothing with a disk sander.**

passes with a concave cutter and two passes with a convex cutter.

The side casings are cut square at the top and surmounted by an architrave head casing, sometimes called a cabinet head. Architrave head casings range from straight and simple to ornate arched and gabled designs, all of which represent the entablature of classical architecture. In contemporary work, where mitered casings prevail, even a simple architrave head casing has an air of classic elegance.

I fitted four of the five new doors with a straight head casing. This casing consists of ¾-in. by 5¼-in. pine, with a simple shaper-cut bead applied to the bottom edges and a standard 1¼-in. pine bed molding applied across the tops. I returned the ends of the beads using a sharp block plane and sandpaper. The bed moldings have full mitered returns, glued and bradded in place.

The head casing on the leaded-glass door takes the classic shape of a broken pediment and urn, a detail that flows gracefully below the arch of the curved ceiling directly above (photo right). A 1x backplate is bandsawn to the pattern and, like the straight head cases, has a bead applied to the bottom and bed moldings applied and returned on the top.

A glass-top door. **The master-bedroom door is fitted with a two leaded-glass top panels, which help illuminate the hall without sacrificing privacy in the bedroom.**

Photo: Bruce Greenlaw

Next, I turned the urn on the lathe in a simple, classic profile and then bisected it on the bandsaw. The head case is coped against the closet head case on the left-hand side.

The urn was installed last. I simply put several daubs of silicone adhesive on the back and pressed it into place.

Hanging and finishing—The doors were trimmed to fit their openings with a hand plane and beveled at a 3° degree angle on the lock stiles so the doors would swing clear of the jambs. I cut hinge mortises in the new doors to match the existing jamb mortises, and then hung the new doors on the old hinges. Rather than reuse the old passage locksets, we installed new lever locksets.

After covering everything with an oil-based primer, we painted the hall ceiling a standard latex flat white, the walls a latex semigloss sky blue and the woodwork a semigloss ivory oil enamel. The transformation of the hall is very satisfying, from cheap and dark to a pleasing harmony of crisp, bright details. □

Joseph Beals is a designer and builder in Marshfield, Mass. Photos by author except where noted.

Production-Line Jamb Setting and Door Hanging

Time-saving techniques from the tracts

by Larry Haun

I went to work as a carpenter in 1949. A craftsman in white overalls taught me the trade I practice to this day, from foundation to finish work. In those early days, my crew and I would build two or three houses a year.

But post World War II America was booming. Good jobs with good wages were available to anyone who wanted to work. Veterans could take advantage of home loans under the G. I. Bill. Literally hundreds of thousands of people needed, wanted and could afford to own a new home. Instead of building one home at a time, carpenters needed to build 5,000 at a time, and as a result, we had to come up with more efficient construction methods.

And that's just what we did. It wasn't long before we were constructing a house in days instead of months. But even though these new methods were considerably faster than the old tried-and-true procedures, they proved to be just as effective.

Everything seemed to change. Hammers gave way to nail guns. Power tools became commonplace. And most important, housing production became an assembly-line process. Those who framed the walls no longer cut and stacked the roof. Those who set the jambs no longer hung the doors.

The new assembly-line methods generated an incredible increase in production. In 1950, we were expected to hang a door an hour and eight doors a day. But in 1953, a friend of mine, a door-hanging specialist, was hanging 80 to 120 doors a day with assistance from a helper. Not only was he fast, he was accurate; the quality of his work far exceeded that of carpenters who hung only a few doors a year.

A clipped-trimmer door jamb stands in the foreground of southern California's Local 409 instructional framing project. Note how the header is just below the top plate, eliminating the need for cripples. The trimmers have been plumbed, straightened and clipped to the king studs, and the jamb has been affixed to the trimmers without shims. The portion of the sole plate that used to be between the trimmers has been removed and installed as drywall backing above the head jamb.

Clipping the trimmers—Traditional methods of setting door jambs are quite effective, yet they require considerable measuring, sawing, nailing and shimming. A faster, but equally effective way to build door jambs is based on a technique called *clipping*. Clipping eliminates the need for shims—nails alone hold everything securely in place. Clipping even eliminates the need for door cripples (photo above), except for jambs in which pre-hung doors are being used.

Because you already know the stud length, there is no need to go from opening to opening, measuring each one. The length of each door header is equal to the width of the door plus 5 in. To set a door jamb using the clipping method, begin by pre-cutting all the necessary trimmers. Cut the trimmers about 1/16 in. oversize to ensure a snug fit. The next step is to put a trimmer under each end of every header. Halfway down its length, secure each trimmer to its king stud with a 16d nail.

Using a 6-ft. level, begin to plumb the trimmer. The level indicates which end of the trimmer doesn't have to move. Toenail that end into the header or bottom plate with one 8d nail in the center. No other face nail is necessary. Then use a straight-claw hammer to pull the un-nailed trimmer end out from the king stud. Once the trimmer is plumb, toenail this end with 8d nails.

Even though the trimmer is now roughly plumb, it will probably have a bow in it. Now it's time to straighten the trimmer (this part of the job eliminates shims). Hold the 6-ft. level on the trimmer (top left photo, facing page) and use the hammer's claws to lever the trimmer away from the king stud so that it's flush with the edge of the level. The 16d nail in the center of the trimmer temporarily holds it straight.

At this point, clip the trimmer to the king stud. Begin by driving a 6d or 8d nail partway into either the trimmer or king stud. Bend this nail back onto the other upright. Then drive and bend a second nail over the head of the first (detail photo, facing page). An experienced jambsetter can drive and bend this nail with one swing. Install three clips per side. This holds the trimmer true for the life of the building. Some jambsetters use a wide staple in place of regular nails to tie the trimmer to the king stud. This is a good method that further simplifies the process.

Use a "spreader gauge" to plumb the second trimmer. Assemble a door jamb, measure the width at the head from outside to outside and cut a 1x4 to this length. Place this spreader gauge up against the already-plumbed trimmer at the top and pull the unset trimmer against the gauge. Secure the second trimmer to the header with 8d nails. Do the same at the bottom. Then straighten and clip this trimmer just like you did the first one.

Now you are ready to cut out the bottom plate. To do this I use a worm-drive saw fitted with an arbor extension that allows me to make a cut that's flush with the saw's base. The one I've got is called a Close Cut, and it costs about $35 (Western Saw Inc., 1842 West Washington Blvd., Los Angeles, Calif., 90007). The Close Cut has a guard on top,

From *Fine Homebuilding* magazine (April 1989) 53:38-42

Trimmers are fastened with the help of a single 16d nail connecting the trimmer and the king stud (top left photo). Here the author levers the two members apart slightly while keeping an eye on the level bubble and the gap between the trimmer and the level. The friction of the nail will hold the trimmer and king stud in the correct relationship while the two are clipped together. Clamping a level to the head jamb allows you the freedom to hold a ½-in. thick block along the edge of the side jamb, approximating the wall-sheathing thickness (photo above). A pair of 8d nails that have been bent over to clip the trimmer and the king stud can be seen in the photo at left.

but it lacks the retractable guard common to circular saws. Because of this, I use it only in flush-cutting situations where the direction of cut is down and away, and I'm careful never to set the tool down while the blade is still moving. If you haven't got a saw fitted with one of these, a chainsaw or a reciprocating saw will do the job. Set aside the cut-out piece for later use as drywall backing.

Installing the jamb—Pick up the assembled jamb and insert it into the framed opening, making sure that it fits snugly. The side jambs will eventually be nailed directly to the trimmers, with no need for shims.

To level the head jamb, use a short level and clamp it to the jamb (top right photo). This frees your hands to do other things. Now use a small block of wood to represent the thickness of the wall covering. Typically, the block is a piece of ½-in. plywood that represents the most common wall covering—½-in. drywall. Place the gauge against the trimmer in order to hold the jamb out the required distance from the face of the wall.

Next, working from bottom to top on the trimmer, nail in five pairs of 6d finish nails into one side jamb. Keeping the pairs of nails separated by 3 in. prevents the jamb from cupping. Larger nails need to be used in jambs holding heavy doors.

Now check the head jamb for level. If the other side jamb needs to be picked up a bit, do so, and then nail it off. You may have to cut off uneven side jambs with a backsaw. This happens, for example, when you want side jambs to sit directly on hardwood flooring. If you are using carpet or vinyl, a small gap under a side jamb will be covered by the finish flooring material.

At this point, retrieve the bottom plate. This plate is the exact length needed to fit on top of the door jamb, where it now becomes drywall backing. It is fixed in place by toenailing through the backing into the trimmer. The backing eliminates the need for header cripples, provides backing for drywall nailing, and makes the head jamb extra secure.

The last step is to cross-sight the side jambs to ensure that they are parallel to each other. If cross-sighting is not done accurately, the door will hang improperly. Cross-sight by eye instead of using the time-consuming method of holding a string diagonally from corner to corner. Stand along the wall and sight along the side jambs to see if they line up with each other (drawing next page). If they don't line up, tap the out-of-line jamb into place by hammering on the bottom plate of the wall until both sides line up. If the building's frame has been properly plumbed and lined, very little correction will be needed to ensure that the jamb sides are in perfect cross-sight.

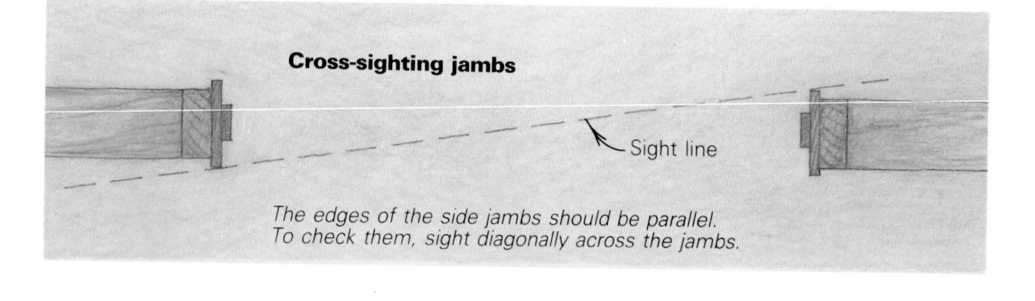

A door hanger's workbench contains all the tools necessary to do the various parts of the job, along with places to support the door while it's being worked (photo above). Here Royal Schieffer uses a ½-in. drill and a lockset jig to make short work of the holes needed for locksets and dead bolts. Just below the lockset jig are two registration marks drawn in pencil on the bench rail for locating the jig's positions. The hollow in the center of the bench allows one person to stand inside it and easily lift it from job to job.

Once you have accurately cross-sighted the side jambs, the door jamb is completely set and ready for its door. The entire jamb-setting operation takes only five or six minutes.

The door hanger's bench—The door hanger's most prized possession is his bench, an ingenious work station that is easy to carry from one job to the next (photo left). While no two benches are exactly alike, they will share certain features. Note how the four corners are covered with carpet to protect the door when it is laid flat. Dowels slide through the legs of the bench to support doors of different widths while they recline on edge. A metal hook at one end of the bench keeps the door from falling sideways, and the corners of the bench are wrapped down the sides with carpet to protect the doors where they lean against the bench.

Inside the bench are bins that contain a router, circular saw, electric plane, electric screwdriver, ½-in. drill and lock jigs. Every tool needed to hang a door is close at hand. And on one side of the bench there are multiple electrical outlets to service all the power tools, eliminating time-consuming plugging and unplugging.

Hinge template—Once the drywall is on the walls, the first step in doorhanging is to rout out the hinge gains (mortises) on the jambs. Production door hangers like Royal Schieffer do this with a hinge template guide (bottom photo, facing page) and a router. One-piece hinge templates can be purchased (see *FHB* #31, pp. 28-31), but door hangers prefer to make their own. Before it can be used, the commercial template has to be secured to the side jamb or door edge with two small pins or nails. Royal's homemade template eliminates this step because it can easily be held in place on the jamb by foot, or on the door edge by hand. When routing hinge gains on the jamb, the template's metal tabs register against the door stop to align it in the horizontal position. The template is shoved against the head jamb, where a round-head screw driven into the top of the template's body acts as a spacer to hold the template ⅛ in. away from the jamb. The spacer gives the door proper clearance between itself and the head jamb.

To rout the hinge gains on the jamb, Royal simply holds the template in place with his foot and runs the router bit around in the hinge guide. Royal uses a ¾-hp router fitted with a collar that rides along the inside edges of the template. The two-flute ½-in. carbide bits that he prefers are made by Paso Robles Carbide (731 C Paso Robles St., Paso Robles, Calif. 93446). Remember that a router is a high-speed power tool and throws out wood chips. Always protect yourself against eye injuries.

Interior doors require two 3½-in. butt hinges. The top of the top hinge is 7 in. from the head jamb. The bottom of the bottom hinge is 11 in. from the bottom of the door. If

To protect fragile door skins, Royal Schieffer's saw has a piece of plastic laminate contact-cemented to its base. The bent end of the metal flange to the left of the blade holds the wood fibers in place during the cut.

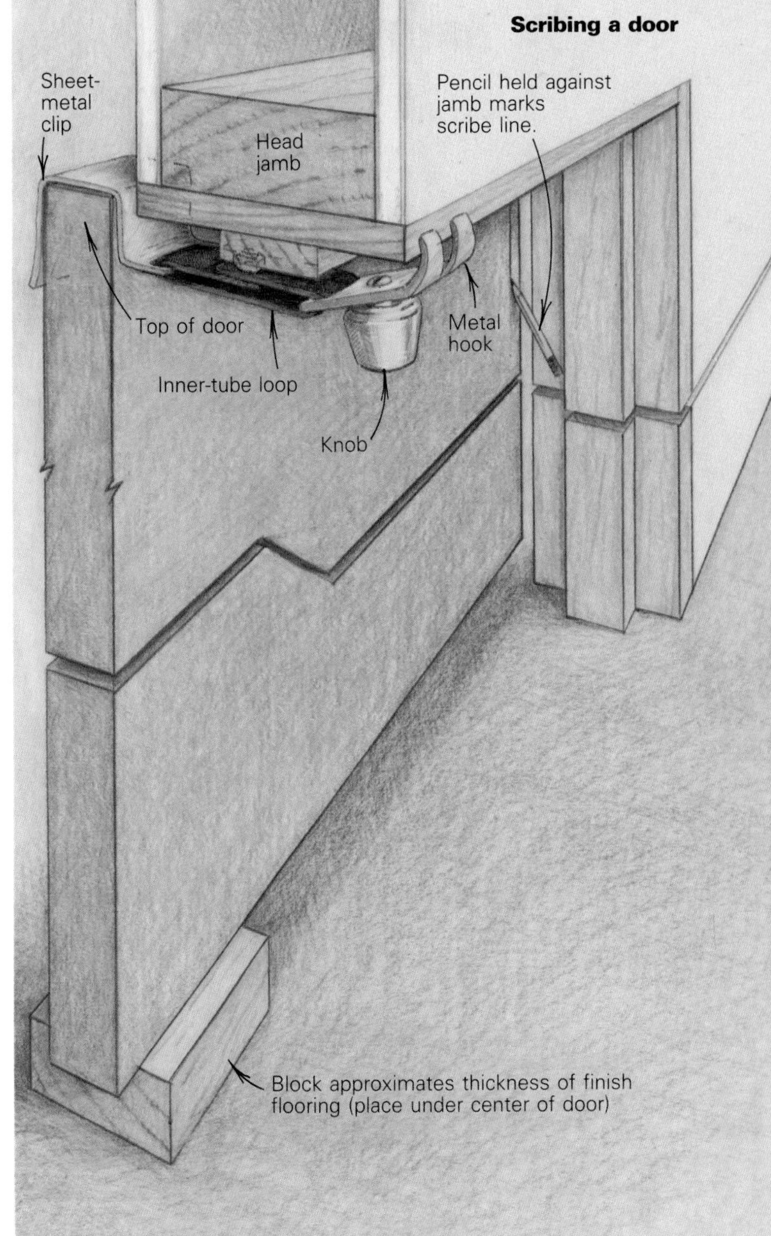

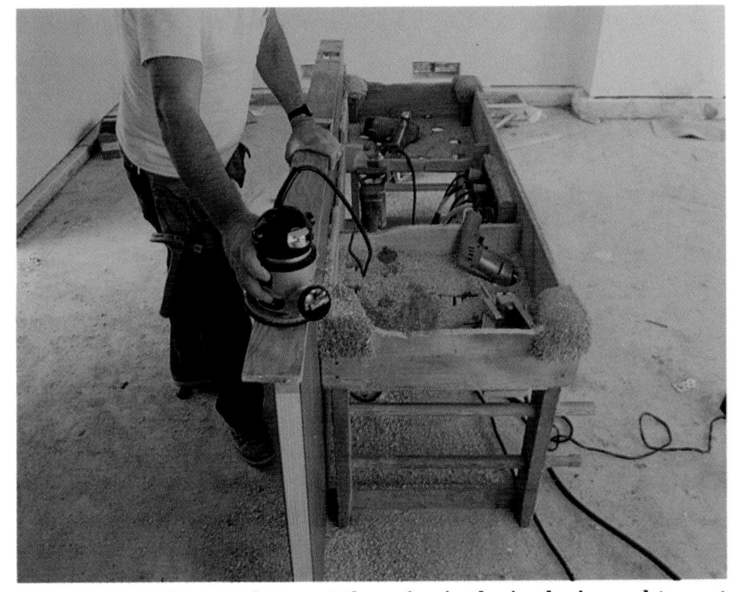

The same template used to rout the gains in the jambs is used to rout the gains in the edge of the door. The end of the template is flush with the end of the door. Note the template's round-head screw: it ensures ⅛-in. clearance between the door and the head jamb.

called for, a third hinge is centered between them. Heavy doors need three 4-in. butt hinges. Once the jambs are routed, it's time to fit the doors.

Fitting a door—Experienced carpenters usually don't need to check the blueprints to see which way a door swings. For example, a bedroom door most often swings in. Check the location of the electric switch. When the door is open, you must have easy access to this switch. If you have any doubts, check the plans and then make a mark on the floor indicating which way the door opens.

At this point set the door in place in the opening in preparation for scribing it to fit. The door is set on a block prior to scribing. Interior doors typically are lifted ½ in. off the floor if vinyl will be the finish flooring, ¾ in. for hardwood flooring, and 1½ in. for carpet. Many door hangers carry a 2x4 block that has been cut into ½-in., ¾-in. and 1½-in. steps for just this purpose.

Royal holds the door in place with a homemade door anchor placed over the top of the door and hooked on the inside of the jamb (drawing above). This tool isn't on the market, but it's easy to make and it's indispensible to a door hanger. It holds the door firmly in place, freeing your hands to complete the scribing process.

The easiest way to scribe a door is with a short, round pencil or a flat carpenter's pencil. Hold the pencil on the jamb and make a mark on both sides and on the head of the door. Unless your door has stiles that extend beyond the bottom rail, there is no need to scribe the lower end of the door. The door can be shortened by removing the excess from the top, requiring only one cut, not two. Mark the hinge location on both the jamb and the door and place the door flat on the bench.

Cross-cutting a door—Cutting across the vertical grain of a door has always been a touchy process. Done incorrectly a saw blade can break slivers loose from the door's veneer, resulting in a ragged cut. One way to avoid "tearout" is to lay a straightedge across the door and score the cutline with a sharp knife. This works quite well, but it's time-consuming. Another way to make the cut is with a saw fence we call a "shoot board." This is a straightedge with a fence screwed to it. The distance between the straightedge and the fence is exactly equal to the distance between the left side of the saw blade and the left edge of the saw's base. Clamp the straightedge to the cutline. It holds down the veneer as the fence guides the cut. A third method is to modify a saw with a homemade flange that is affixed to the top of the saw's base (photo above left). The flange is adjacent to the saw blade, and it rides lightly on the door during a cut, preventing the veneer from lifting. It really works.

Running a metal saw base over a door often damages or scratches the finished wood. A way to avoid such damage is to glue a

Drawings: Christopher Clapp

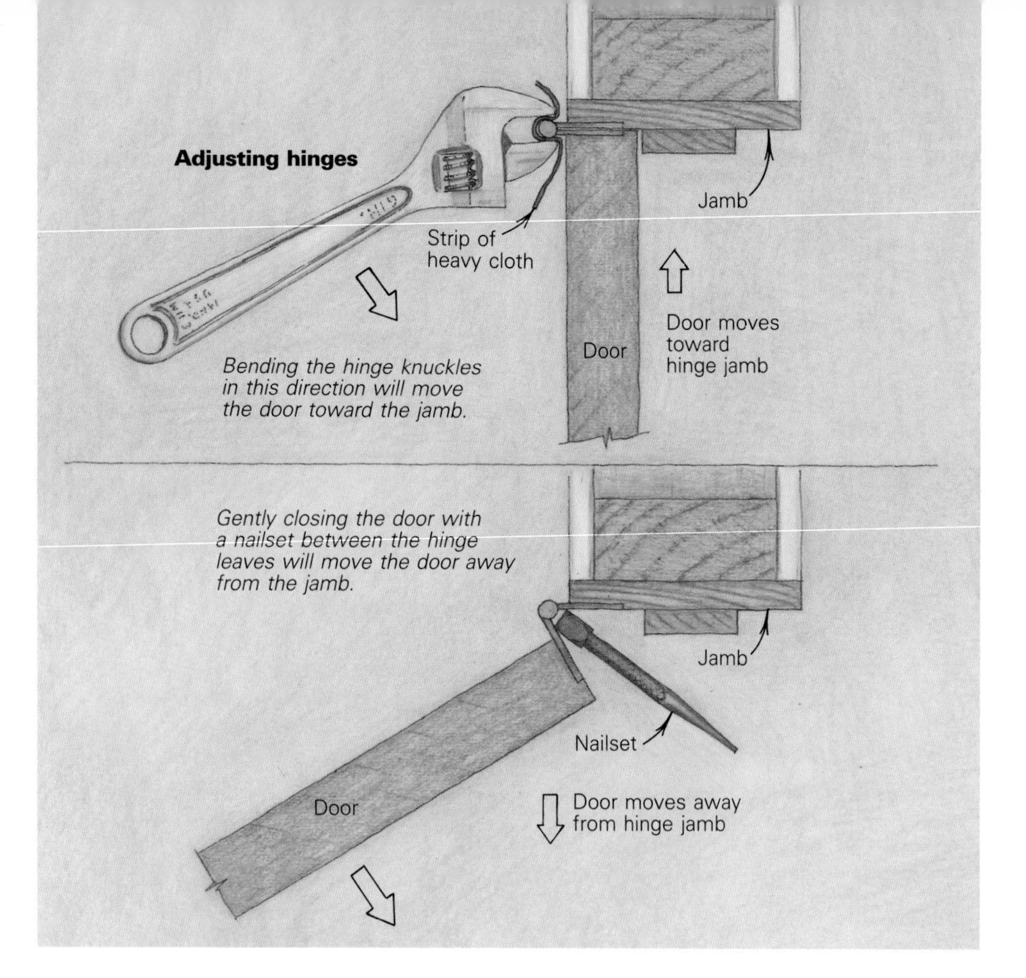

piece of plastic laminate to the base of your door-hanging saw. It's also a good idea to attach laminate to the fence of the electric plane. Rubbing a little paraffin on the laminate makes these tools float over the work.

When you cut the top of the door, bevel it about 2° to the inside. Do this for two reasons. First, it eliminates problems that occur from paint buildup over the years. Second, once the door is hung, you may have to take a bit off the top in places to ensure proper clearance all the way across. It's easy to do this with a block plane without removing the door. All you have to do is take a bit off the high edge—it's unnecessary to plane the entire head. The bevel gives plenty of clearance on the stop side. Hangers refer to this process as "fine-tuning the door."

At the bench—Once the door has been crosscut, it's time to place it on edge at the bench. The doorhanger now fits and bevels each side with a power plane. Doorhangers typically set the fence of the plane at 3° to achieve the proper bevel. Everyone knows that it's necessary to bevel the lock side. But it's also important to bevel the hinge side to prevent difficulties with paint buildup. And what's more, if the jamb is cupped, you won't have problems with hinge bind.

The amount of clearance between the door and the jamb depends on the region in which you live. In most situations, leave a little less than a ⅛-in. gap at the top and on the sides. You'll automatically do this by removing the scribe line with the plane set at the 3° bevel. If the house is in a humid climate where the door is going to absorb moisture, take off another ¹⁄₁₆ in. to keep it from sticking when it expands. In dry climates, the clearance can be less.

As you cut to the scribe line, keep a thumb on the front lever of the plane. Watch ahead of the plane, raising or lowering the cutter to the required depth. If there is an excess of material to be removed, you may have to make two or three passes with the plane.

Now turn the plane so that its cutter is at a 45° angle to one of the door's edges and put a slight chamfer on the edge. Done properly, all the door's edges can be dressed with the plane to make them look like factory edges. A pass or two with sandpaper puts the finishing touch to this stage of door hanging.

Routing for hinges—The next step is to rout the hinge gains on the door and install the hinges. Hold the template flush with the top of the door and cut the hinge gain with the router (bottom photo, previous page). Contemporary hinges have rounded corners so there is no need to chisel out the corners of the gains. Place the assembled hinges in position and drive in the screws with an electric screwdriver. Pilot holes are unecessary unless you're installing a hardwood door. With a little practice, you should be able to install the hinge perfectly (all of this is done right at the bench).

When you're finished on the hinge side, flip the door over to the lock side. The position of the locks—36 in. from the floor for the door knob, 42 in. for the dead bolt—is marked on the bench. There's no need to measure each door individually. Simply register the jig to the marks, clamp it in place (photo, p. 106) and bore the holes for the cylinder and the latch.

Hanging a door—Like many other carpentry techniques, door-hanging gets easier with practice. An experienced door hanger doesn't have to pull the pins and split the hinge leaves—even with heavy, solid-core doors. A typical installation takes about five minutes.

Take the door, along with your battery-powered screw gun, to the opening. Place the bottom of the door on the end of your foot so that you can insert the top hinge in the jamb gain. Once the door is in position, secure the top hinge into the jamb with one screw. Next, swing the door back parallel with the wall. Put your toe against the bottom edge, push the door plumb and, like magic, the other hinge should drop into its gain. Now it's just a matter of driving home the remaining screws.

Before doing anything else, check the door for fit. The door should fit almost perfectly, with little need for fine-tuning. Because the top and sides of the door are beveled, it's easy to shave the door a little with a block plane to make the clearance the same all the way around the door.

Sometimes the door will be the correct size, but the hinge keeps it too close to one side. Carpenters used to correct this by putting a piece of cardboard behind the hinge, but nowadays there's an easier way.

Over the years, door hangers have noticed that hinges vary from case to case. Hinges from one box may hold the door just a little closer to the jamb side than hinges from the next box. To adjust for this you can spring the hinges slightly. To move the door away from the jamb, stick the butt end of a nailset in the hinge and close the door gently (bottom left drawing). This opens the hinge a bit. To move the door closer to the jamb, use an adjustable wrench on the hinge knuckles (top left drawing). To keep from marring a hinge, use a piece of heavy cloth with the wrench. At times, I've heard criticism that this isn't craftsmanlike. Personally, I think the problem isn't with the craftsman, but with the variations in hinges. If your client is providing you with first-quality hinges, this problem shouldn't arise.

With the hinges adjusted, it's time to make one last check by cross-sighting the jambs. Other tradesmen can sometimes bump a wall and knock jamb sides out of parallel. The finished door, when closed, should be flush with the jamb on all three sides. If it's not it can usually be corrected by moving the wall a little. You can do this by placing a block against the drywall at the bottom plate and tapping the block until the door is flush with the jamb. To make sure that the wall stays in the desired position, drive a small wedge under the bottom plate. The weight of the building will keep everything in place. The door is now ready for a lock, stops and casing. □

Larry Haun lives in Los Angeles and is a member of local 409, where he teaches carpentry in the apprenticeship program.

Pocket Doors

Should you buy a kit or build your own?

by Kevin Ireton

When I asked one architect about pocket doors—doors that slide into a wall, rather than swing on hinges—he said he avoided them like the plague. Another told me he used them only as a last resort. One builder simply said, "They're stupid. Don't use them." Modern-day pocket doors have a reputation for creating flimsy walls on either side of the pocket, for using lightweight hardware that's easily misaligned and rollers that jump off the track, and for needing repairs that are impossible without tearing open the wall.

But blaming pocket doors for these problems is like blaming a circular saw for not cutting straight. Having installed a few pocket doors myself, and having recently spoken to builders all over the country about pocket doors, I've learned that, properly installed, some commercially available frame kits work just fine. I also discovered a great system for building your own pocket-door frames.

Sliding versus swinging—To swing 180°, a standard 2-ft. 8-in. door needs over 10 sq. ft. of clear floor space. A typical house—three bedrooms, two baths—might have nine or ten swinging doors whose collective door swings lay claim to 100 sq. ft. Half of that figure represents the space directly in front of the door, which has to remain clear anyway. But the other half is usable space—50 sq. ft. of it—that could be reclaimed through the use of pocket doors.

Pocket doors are commonly used for two specific reasons: either because the space and traffic pattern demand it, such as in a half-bath off a hallway where it would be awkward to have a swinging door; or because you want the option of occasionally closing off a room without having to sacrifice space in return. An example of the latter would be pocket doors used between kitchens and dining rooms. Such doors are going to be in the pocket 90% of the time, but when you want to hide the dirty dishes after Thanksgiving dinner, you can pull them shut.

One advantage of pocket doors is that the two sides can be painted different colors, or even milled differently, to match the rooms they face. A swinging door, on the other hand, shows both sides to the room it swings into—one side when the door is open, the other side when it's closed.

Structurally, pocket doors have the disadvantage of requiring a header that's twice as long as that of a swinging door. The problem is magnified with converging pocket doors—a pair of pocket doors that slide toward each other. For example, if you have a 6-ft. opening between two rooms and you're debating double swinging doors versus converging pocket doors, the latter will require a header that's nearly 13 ft. long. Another disadvantage of pocket doors is that if they're used in a 2x4 wall, you don't have room for an electrical outlet or switch in the area of the pocket. None of these problems is insurmountable, though, and if you don't have room to swing a door, sliding it into a wall may be your best option.

Pocket-door frame kits—Two types of pocket-door frames are available off the shelf, for use in 2x4 walls (though they can be furred-out for thicker walls). One type comes sized for specific doors and includes two preassembled wall sections and a header/track assembly. The wall sections are ladder-like affairs made of 1x stock, with two vertical pieces connected by three or four horizontals.

Nobody I've talked to likes these kinds of frames. The tracks are usually light-gauge steel with single-rollers on the door hangers, and they're rated only for 50-lb. or 80-lb. doors. (Although most standard interior doors weigh less than 50 lb., the weight rating offers some indication of how easily the door moves on the rollers.) The lumber used in these units is of poor quality, seldom straight, and the installed wall sections are flimsy. Even one of the manufacturers I talked to (who also makes the other kind of door frame) told me it wasn't a good frame.

The better frames are "universal" pocket-door frames, available now from a number of

Universal-type pocket-door frame. **The header/track assembly (top left) is nailed to the end studs, not to the framing header, which allows the header to sag a bit without bending the track. Wrapped with steel on three sides, the pocket-door studs are screwed or nailed to the sides of the header/track assembly at the top (top right), and at the bottom, slip over prongs on a metal bracket nailed to the floor (bottom right). A rubber bumper, attached inside the pocket (bottom left), limits the travel of the door and prevents it from banging into the end stud.**

From *Fine Homebuilding* magazine (June 1989) 54:63-67

companies (see chart, facing page). They cost around $40 or $50 (not including the door) and can accommodate most doors that are 2 ft. to 3 ft. wide and 6-ft. 8 in. high (some companies offer frames or extension hardware for doors up to 5 ft. wide and 7 ft. high). The frame kit includes a header/track assembly, two door hangers and four 1x2 studs. The studs are wrapped with steel on three sides to stiffen the wall and prevent nails and screws from penetrating into the pocket (top right photo, previous page). The tops of the studs should be nailed into the side of the header/track assembly, but the bottoms slip over the prongs of a metal bracket nailed to the floor (photos previous page, top and bottom right), which lets the floor sag a bit without the studs pulling the track down and bending it.

The door hangers supplied with these frames have at least three nylon rollers (some have four), and the tracks are shaped so that the hangers have to be slipped over the end of the tracks, which makes them virtually "jump-proof." Most of the universal frames now come with a two-piece hanger. The roller unit is one piece, and it slips into the track with a threaded stud that hangs down. The other piece is a bracket that you attach to the top of the door.

When you attach the brackets, be sure to place them far enough from the end of the door that both screws go into the top rail, not into the end grain of the stile. After the brackets are attached, the door is lifted into place, the bracket engages the stud and is usually locked by a little plastic gate that you swing shut. The door can be adjusted by screwing the stud in or out of the roller housing with a special wrench included with the kits.

One of the criticisms of pocket doors is that you have to tear out the wall if something goes wrong, but some of the manufacturers (Johnson, Sterling and Acme, for instance) have designed their tracks so that you can remove them for repair or replacement without disturbing the wall. After taking down the door, you remove the mounting screws in the section of track above the doorway. The remaining screws, inaccessible in the pocket, sit in keyhole-shaped slots; simply pull the track toward you a fraction and the track will slip over the screws. Before installing the track, check that these screws are not too tight or you'll never get the track out later.

Most of the universal kits are rated for 100-lb. or 125-lb. doors. Some companies offer heavier duty hardware if you need it. Acme has a universal frame rated for 250-lb. doors.

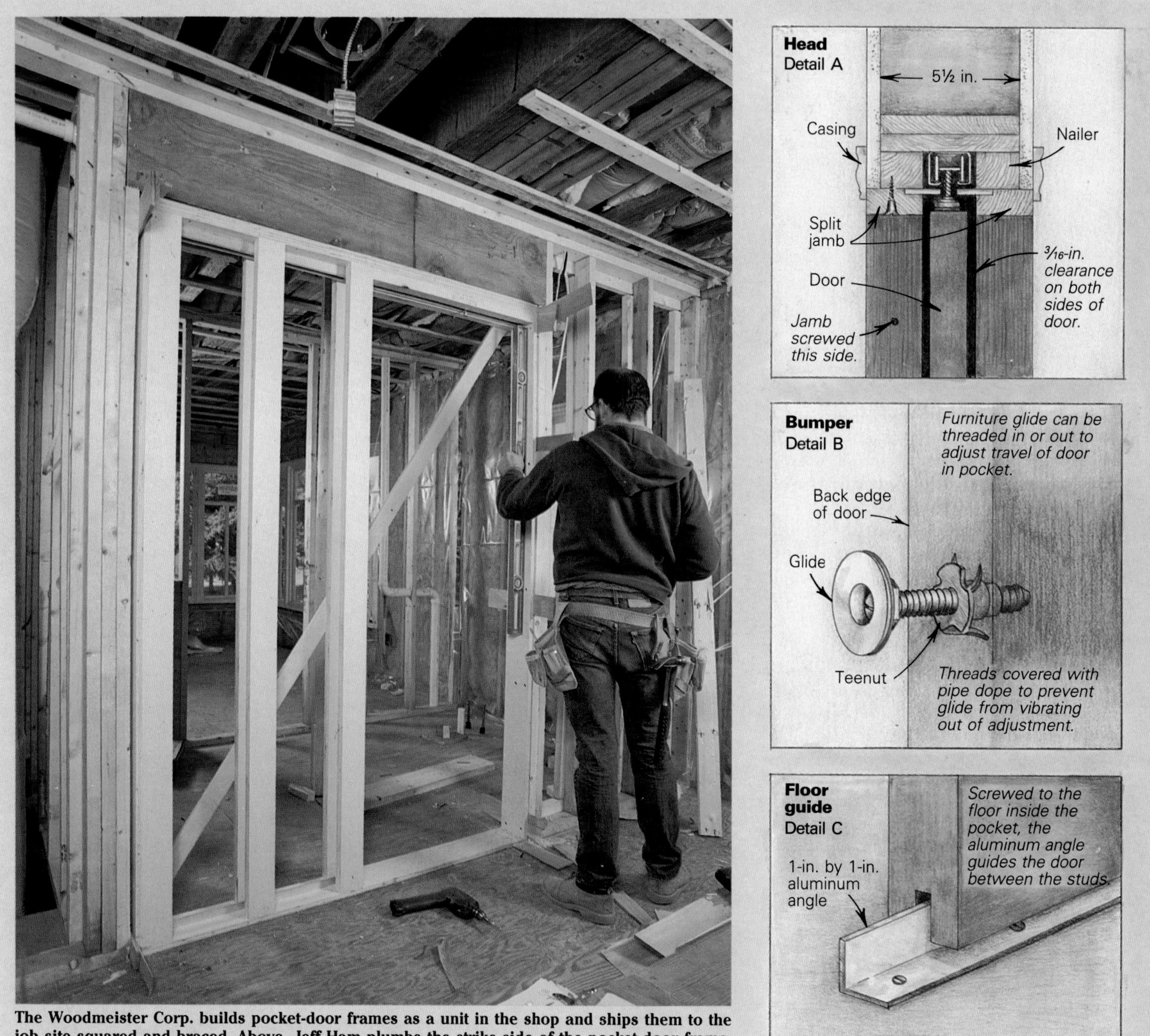

The Woodmeister Corp. builds pocket-door frames as a unit in the shop and ships them to the job site squared and braced. Above, Jeff Ham plumbs the strike-side of the pocket-door frame. The plywood plate running across the doorway will be cut out after the frame installation.

Drawings: Michael Mandarano

And Johnson Products has a set of optional ball-bearing hangers rated for 200-lb. doors. With a 125-lb. door and Johnson's standard rollers, the folks at Johnson say it takes between 6 lb. and 7 lb. of pull to slide the door out of the pocket. With a 200-lb. door and Johnson's ball-bearing hangers, it only takes 3 lb. of pull.

When I asked if the ball-bearing hangers would hold up any better, the folks at Johnson said that they didn't think so. They've cycle-tested their standard rollers using a 150-lb. door, and after 100,000 trips in and out of the pocket, all they got for their trouble was a black stripe around the white nylon wheel.

Frame installation—Installing a universal pocket-door frame is not difficult, but you have to be careful. The trouble is that pocket doors must be installed while you're framing the house, before drywall, and in a mood to work quickly rather than carefully. The manufacturer's instructions explain the basic installation, but here are some things to keep in mind.

The track should be nailed between the trimmer studs on either side of the rough opening. The track is not attached to the framing header, so there should be a space between them (top left photo, p. 109). This allows the header to sag a bit without bending the track.

Before you install the studs, site down their length. Three out of four studs in the last frame I bought had a ¼-in. bow in them. You can straighten them with your hands, which indicates how easily they can be knocked out of alignment. The studs should be carefully

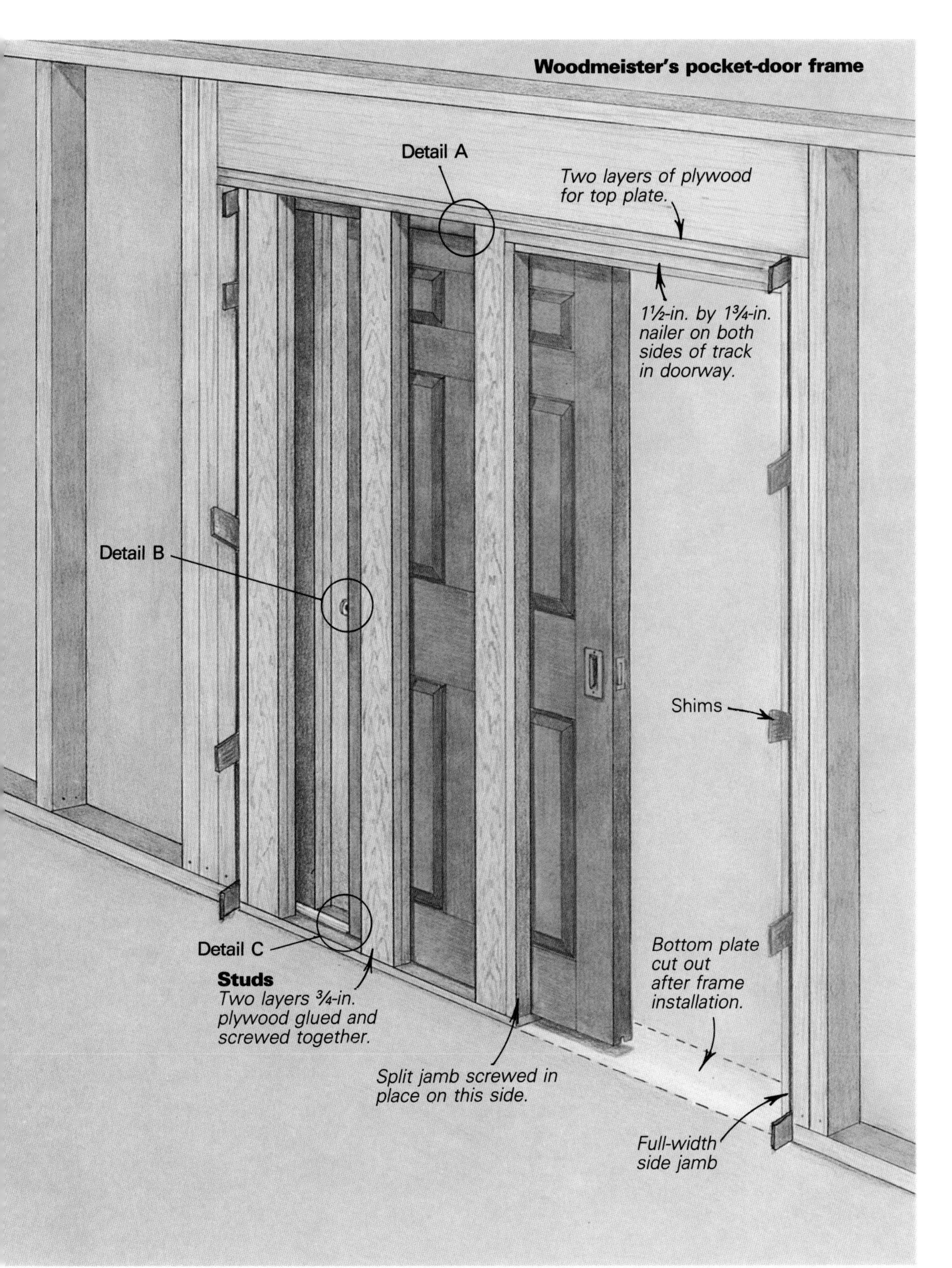

Manufacturers of pocket-door frame kits and hardware

Manufacturer	Frame kits	Tracks and rollers	Pulls and latches
Acme General Corp. Div. of the Stanley Works, 300 East Arrow Highway, San Dimas, Calif. 91773	●	●	●
Baldwin Hardware Corp. 841 E. Wyomissing Blvd., Reading, Pa. 19611			●
G-U Hardware, Inc. 11761 Rock Landing Dr., Suite M6, Newport News, Va. 23606		●	
Grant Hardware Co. High St., West Nyack, N. Y. 10994-0600		●	●
Hafele America Co. 3901 Cheyenne Dr., P.O. Box 4000, Archdale, N. C. 27263		●	●
Iseo Locks Inc. 260 Lambert St., Suite K, Oxnard, Calif. 93030			●
L. E. Johnson Products, Inc. 2100 Sterling Ave., P.O. Box 1126, Elkhart, Ind. 46515	●	●	●
Lawrence Brothers, Inc. 2 First Ave., P. O. Box 538, Sterling, Il. 61081	●	●	●
Merit Metal Products Corp. 242 Valley Rd., Warrington, Pa. 18976			●
National Manufacturing Co. 1 First Ave., Sterling, Il. 61081	●	●	●
Quality Hardware Manufacturing Co., Inc. 12705 S. Daphne Ave., Hawthorne, Calif. 90251			●
John Sterling Corp. 11600 Sterling Parkway, Box 469, Richmond, Il. 60071-0469	●	●	●
Wing Industries, Inc. 1199 Plano Rd., Suite 110, Dallas, Tex. 75238		●	

plumbed in both directions, especially the two at the end of the pocket that will support the split jamb. It's a pain in the neck to shim the split jamb to make it plumb.

After you've installed the frame, insert a couple of temporary spacers (1x stock, cut to the width of the pocket) horizontally in the pocket to prevent the studs from bowing inward while the drywall is hung. Even with the spacers in place, though, whoever is hanging the drywall should be careful not to bear down so hard on the screwgun as to knock the studs out of alignment. Shorter screws should also be used.

Be sure the full-width side jamb that the door closes against is plumb. If this piece and the split jamb opposite it are not both plumb, then you can't adjust the door so that it both closes properly and rests flush with the split jamb when open.

To allow access to the hangers (in case you have to remove the door later), you'll need to screw one side of the head jamb in place rather than to nail it. This means you should put the side jambs up first; otherwise they'll trap the head jamb. Also, remember not to nail the casing into the head jamb that's screwed in place.

Make sure that the door you use is straight and flat, and that it's sealed on all sides, including the bottom. If the door warps, it will hit the studs in the pocket and won't open or close. Remember, too, that more than one pocket door has gotten scratched or nailed in place by a carpenter who forgot (or didn't know) that he was nailing into a thin wall. It won't hurt to remind subcontractors about the pocket door, either.

Building your own—If you have to install a pocket door in a 2x4 wall, you might as well use a kit. Although I've spoken to some builders and architects who think otherwise, most agree that the kits work pretty well. On the other hand, if you can afford to beef up the finished wall thickness to 6½ in. (2x6 framing), you're probably better off buying the hardware and building your own frame. This is especially true if you're using bigger than average doors. Besides letting you construct stiffer walls on either side of the pocket, a 2x6 wall gives you enough room to install shallow electrical boxes.

The best system I've seen for building your own pocket-door frame is the one worked out by the people at the Woodmeister Corp. (drawing previous page), an architectural woodworking company in Worcester, Mass. They start with heavy-duty tracks and rollers (Woodmeister uses Lawrence hardware, but Grant and Acme also make some pretty rugged stuff). Then they build the frame using ¾-in. cabinet-grade veneer-core plywood (poplar or birch) rather than solid lumber, because it's more stable.

The entire frame is assembled in the shop. It's built as a unit to fit the pocket-door's rough opening (photo, p. 110). The top and bottom plates run the full length of the rough opening, with a 5½-in. wide stud at each end. The bottom plate, which is continuous across the door opening, is cut out later. The studs are made up of two layers of plywood, each 3½ in. wide. Before the layers are glued and nailed together, the fabricator sights their length and orients the pieces so that any bow in one is countered by the bow in the other.

In addition to the overhead track, Woodmeister uses a continuous floor guide inside the pocket that centers the door and prevents it from hitting the studs. A piece of 1-in. by 1-in. aluminum angle, running the length of the pocket and extending about ¼ in. past the last stud, is screwed to the bottom plate (bottom drawing, p. 110). A corresponding groove is routed in the bottom edge of the door to receive the angle. After assembling the frame, the fabricator squares it up and screws a full-length diagonal brace across it before shipping it to the job site.

Installation is straightforward. As with the off-the-shelf frames, the top is not attached to the header of the rough opening, and the important points are that the track is level and the studs plumb in both directions. Woodmeister's installers shim the frames and screw them in place to avoid the possibility of repeated hammer blows knocking the frame out of alignment.

The standard frame kits supply you with a little rubber bumper to nail on the end stud in the pocket to cushion the blow from the door hitting it (bottom left photo, p. 109). Woodmeister installs a teenut and an adjustable plastic floor glide (like the kind used on the legs of office furniture) in the back edge of the door (middle drawing, p. 110). This cushions the blow of the door but also allows adjustment of the travel of the door into the pocket. Before installing the floor glide, pipe dope is smeared on the threads of the glide to keep it from vibrating loose over time.

When installing the split jamb, the folks at Woodmeister screw one side of both the side and head pieces to the plywood studs. This makes removal of the door easier.

Most people don't use any kind of door stops on a pocket door. The leading edge of the door simply butts into the jamb. Woodmeister has tried cutting a ¼-in. deep rabbet into the full-width side jamb to receive the door edge. But while this looks nicer when the door is closed, it doesn't look as good with the door open. Also, the use of a rabbet or stops can lead to trouble if the door warps at some point down the road.

The edge pull is used to get the door out of the pocket, and the flush pulls, installed on both sides of the door, are used to open the door (photo above). But with this setup, there is no way to latch or lock the door. The unit shown in the top photo combines face pulls, edge pull and privacy lock all in one. To install it, you simply cut a notch in the edge of the door; no mortising is required.

Pulls and latches—During the Victorian period—probably the heyday of the pocket door—some wonderful decorative hardware was available for pocket doors, including ornate recessed pulls for the face of the door and great locksets with edge pulls that popped out when you pushed a button. Unfortunately, no one that I know of is reproducing them, so you'll have to shop the salvage yards if that's what you're after.

Here's an overview of today's options. You can use an edge pull in the edge of the door (a nice one is available from H. B. Ives, A Harrow Co., P. O. Box 1887, New Haven, Conn. 06508) and a pair of flush pulls (available from various companies) on both sides of the door (bottom left photo). Or you can get a lockset that includes flush pulls, edge pull and a privacy latch in one unit (top left photo). The ones I've seen were made by Quality Hardware Manufacturing Co. (see chart on p. 111 for address). There's no mortising with this unit. Instead, you cut a 2¼-in. by 1¾-in. notch in the edge of the door. Such a big bite, though, could affect the integrity of some doors, especially hollow-core doors, and perhaps lead to warping. In any case, this type of lockset will likely void any warranty on the door.

Some of the companies that make frames (Lawrence and Johnson, for instance) sell latches and face pulls that fit the holes for standard swinging-door locksets. If you want real security, Baldwin, Hafele, Merit and Iseo (see chart for addresses) make case-style locksets available with key cylinders. □

Kevin Ireton is an associate editor of Fine Homebuilding.

About old pocket doors

by James Boorstein

I cannot put a date on the earliest use of pocket doors in this country, but I know they were used in the 18th century. As American architecture evolved beyond its rustic and purely functional roots and grew increasingly grand, large pocket doors became quite common. Possibly it was an easy way for some builders to deal with huge doors without having to use massive hinges and heavy framing. Pocket doors were used to separate the more public rooms of the house—the parlor, library and dining room—from each other. Rarely were they used on the upper floors to separate bedrooms.

Most of the early domestic pocket doors had wheels on the bottom and rode in a track on the floor. Not until the middle of the 19th century were overhead tracks and rollers available. The switch to an overhead system was probably made as hardware technology advanced. The overhead track was out of sight and was less susceptible to problems resulting from dirt in the track and from the floor settling.

The older, more traditional pocket doors were almost always used in pairs (photo below). Each door rolls on two wheels that are either solid or spoked and 4 in. to 5 in. in diameter. Most are seated in a permanent housing called a "sheave," which looks somewhat like a window-sash pulley. Many of the wheels, especially those found in the southern U. S., can be very decorative, even though they aren't seen. Some of the sheaves are just long enough to house the wheel itself, other sheaves are two or three times the diameter of the wheel with a horizontal slot that allows the axle of the wheel to move forward or backward as the wheel rolls and the door moves. I'm not sure of the exact function of this. It certainly reduces the wear of the axle on the housing and perhaps changes the balance of the door, making it easier to move.

The sheave is mortised into the lower rail of the door (drawing below right), shimmed and fastened with wood screws to allow the door to sit level. Only about ¼ in. of the wheel protrudes below the edge of the door. The edge of the wheel is grooved to fit over a ridge in the metal track, which is often bronze or brass, and sometimes steel. The floor track is often surface-mounted, but is sometimes recessed into the finished floor. In many older homes the floor track appears to be raised. This is usually a result of the floor around the track being sanded away over the years. Lowering the track would affect the operation of the door.

To keep the door vertical, two hardwood pegs (often oak) are mortised into the top rail of each door. Even on the finest doors these pegs are fairly crude. They protrude 2 or 3 in. up into a wooden track. The track has two recesses. The top edge of the door extends into the first recess, and the hardwood pegs ride in the second. There must be adequate clearance above both so that the doors can be lifted up and over the floor track for installation and removal. Occasionally the wooden upper tracks were constructed in such a way that they float and actually lift up as the door is being lifted.

Converging pocket doors have a center stop, generally a sturdy piece of cast hardware screwed to the upper track and shaped to receive the leading edge of both doors. This must be removed to take the doors out. The leading edges on a pair of converging pocket doors were often milled to fit into each other, like a shallow tongue and groove.

There is also a stop, or bumper, on the trailing edge of the door so that the door will come to rest at the proper place when it is slid into the pocket. This stop, which is never seen, is often a very crude block of wood attached to the door at approximately the height of the pulls.

It is common to see traditional round door knobs on pocket doors; the stop keeps these doors 4 in. to 6 in. out of the pocket or just less than the width of the rail. Usually the knobs are fixed and act only as pulls to open and close the door. Locks and latches on the old doors vary, but they all work on the same principle: a curved metal bolt arcs out of the lockset in one door and down into a receiving plate on the other door or on the wall. These locks are often operated by a short, decorative key that works from either side of the door. On doors that rest fully recessed into the pocket when open, flush with the jamb, I've seen an array of ingenious pulls and spring-loaded pop-out handles that are flush until called into action.

Making repairs—Occasionally old door pockets lie hidden behind contemporary walls. Often these hidden pockets still house their rolling doors, but getting them out can be difficult. If they have been hidden through several renovations, they may have electrical cables or plumbing lines run right through them. More common problems include warped studs in the pocket, or a warped door. The door could be off its tracks, it could have been inadvertently screwed or nailed in place, the track could have worked loose or there could be any number of other problems. A screw eye and a loop of rope attached at the top and bottom of the door will serve as a temporary handle to coax a stubborn door out of its pocket.

Once the door is out, get a couple of droplights or electric clip-on lights and shine them in the pocket (a flashlight is no substitute). By carefully examining the pocket and employing some common sense, you will probably be able to figure out and correct any problems.

Restoring the operation of a pocket door should take place after any necessary structural repair in the area has been accomplished. But it should be done before any final plastering, as you may need to make holes in the wall to make adjustments inside the pocket. If for some reason the walls can't be disturbed (if there's wallpaper or a mural, for instance, or if the wall is otherwise historically significant), you can remove material from behind the baseboard and gain access to the pocket and the track without damaging the walls above.

Sometimes the door itself needs basic structural repair. The mortises for the wheels on the lower rail of the door frequently weaken the door. A partial or complete patch, with a new and relocated mortise, will solve the problem. The tracks should be straightened and shimmed level. On older doors the wheel bearings may be worn out. The simplest way to repair this is to pad out the mortises and install replacement sheaves (available from Grant Hardware, see chart on p. 111 for address). Usually, they will work on the existing track. On historic pocket doors I have had the worn hardware remachined, which worked well. While I retained the original hardware, however, I also preserved the original slightly stiff operation of the door. A compromise would be to use a better material for the bearing.

James Boorstein is a partner in Traditional Line, *a restoration company in New York City.*

Nine ft. tall and 3 in. thick, these converging pocket doors reside in the Dakota, one of New York City's most famous apartment buildings.

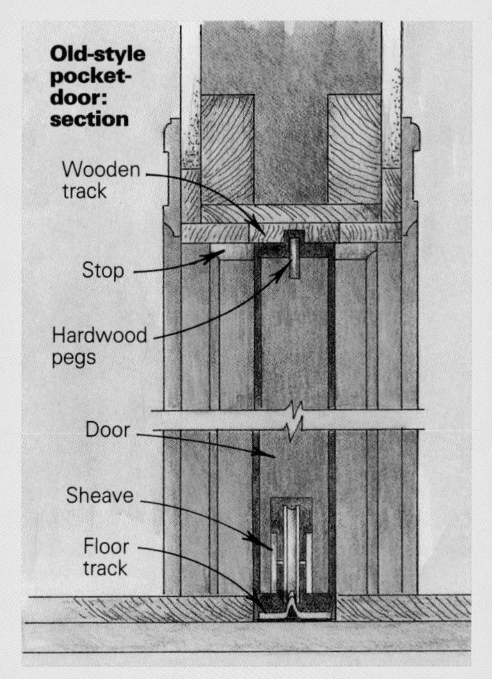

Building Wooden Screen Doors

Durability and aesthetics hinge on sound joinery and steadfast materials

by Stephen Sewall

According to Steven J. Phillips' Old-House Dictionary, the purpose of a screen door is "to allow ventilation but exclude insects." Of course if it can accomplish that gracefully, all the better. During the late 1900s, when screen doors first became popular in this country, most were designed to enrich, or at least complement, the architecture. To that end, a wealth of styles emerged, many of which were featured in pattern books and copied by carpenters. Even many of the mass-produced screen doors in those days were both fetching and functional.

Nowadays, though, the ubiquitous aluminum screen door is the one most likely to appear on someone's doorstep. To be sure, aluminum doors have their place, but all too often they wind up in classical entryways or over finely wrought entry doors, spoiling the intended effect.

A wooden screen door, on the other hand, can be unobtrusive, allowing the main door to show through. Or it can mask the door behind it and serve as an architectural ornament in its own right (photos facing page). As a homebuilder and architectural woodworker, I've built screen doors both ways. And in either case, the basic construction principles are the same.

Screen-door anatomy—In its simplest form, a screen door consists of two stiles, a top rail and a bottom (or kick) rail (drawings, p. 116). Traditionally, most screen doors were built of pine, oak and other domestic woods. But pine is soft, and oak isn't especially stable or weather resistant. Cypress is a good choice, but here in Maine it's hard to find. Though I've used a variety of woods for my screen doors, I prefer Honduras mahogany. It's strong, stable and holds paint and varnish well.

I build my doors 1⅛ in. thick, a compromise between light weight and sturdy construction. Originally, I joined stiles and rails with dowels because it was quick and easy. But after a while, the joints loosened. Screen doors bear the brunt of the weather and they get slammed a lot, especially when they're spring-loaded. Pneumatic closers (the kind you see on most aluminum screen doors) are especially hard on wooden screen doors because they hesitate in mid-swing and cause excessive racking. Nowadays, I use mortise-and-tenon joinery exclusively for my screen doors.

Stiles and top rail are a minimum of 4-in. wide, which makes for a sturdy door and provides plenty of room for locksets and springs. The kick rail is at least 8-in. wide to keep the door from distorting. I cut two narrow tenons instead of one wide tenon in the kick rail so that the mortises don't weaken the stile.

Tenons are 3 in. long and ⅜ in. thick, or one third the thickness of the stock. A ½-in. wide by ⅜-in. deep rabbet along the inside edge of the frame accommodates the screen and screen stop. That's all there is to it.

Making the door—One approach for making this type of a door is to cut the mortises and tenons and glue up the door first, and then rabbet and groove its inside edge with a router to accept the screen and stop (this requires that the rabbets be chiseled square at the corners). But I prefer to rabbet the stiles and rails *before* assembly, extending the rabbets down the full lengths of the stiles. This requires that the ends of the rails be cut to engage the edges of the stiles (top right drawing, p. 116). The cuts are easily made with dado blades on a table saw, and the resulting lap joints strengthen the door.

I start by selecting clear, straight-grained stock with a moisture content of about 10%. After sizing and jointing the stock, I lay out the mortises on the stiles. I stop the bottom mortise about 1-in. short of the bottom end of the stile so that the door can be trimmed without cutting into the mortise and tenon. One inch of stock separates the two mortises for the bottom rail. On the top end, I hold the mortise back about ½ in. On the inboard sides, the mortises stop where the rabbets begin.

I cut the mortises with a horizontal boring machine. I prefer this over a hollow-chisel mortiser in a drill press because the boring machine is faster and more accurate. The only drawback is that it produces mortises with rounded corners. To avoid squaring the ends of the mortises or rounding over the edges of the tenons, I cut the tenons narrower than the mortises by the diameter of the mortising bit, or ⅜ in. That leaves small half-circle hollows on either side of the tenons, which fill up with glue during glueup. I also cut the short rabbet at the mouth of each mortise with the boring machine, squaring its inside corner with a chisel. Mortises can also be cut with a router or chopped out by hand with a mortising chisel.

I cut the tenons on a table saw fitted with a ¾-in. wide dado head set to depth of ⅜ in. I lay the stock flat on the table and push it through carefully with a miter gauge set to 90°. The rip fence on the table saw serves as a stop to determine the length of the tenon. Starting with the shoulder cut and pulling the stock away from the fence with successive cuts, I need to make only five or six passes to cut a 3-in. long tenon.

I readjust the fence to cut the other side of the tenon so that it's ½ in. longer than the first side to compensate for the screen rabbet. After the cheek cuts are completed, I adjust the height of the dado head, flip the rails on edge and trim the tenons to width using the same method. I cut out the space between the double tenons on the bottom rail with a bandsaw. Tenons can be cut a number of other ways, of course, such as with a single-end tenoner, a tenoning jig on a table saw (see *FHB* #36, pp. 39-41), a bandsaw or even by hand with a backsaw.

With the mortises and tenons completed, I cut all the rabbets using either the dado head on the table saw or a shaper, which gives a cleaner cut. The ½-in. width of the rabbet gives ample room to center a ⅛-in. wide by 3⁄16-in. deep (approximately) groove. The screen will be pressed into this groove and held fast with ⅛-in. dowel stock (more on that later).

I cut this groove with a combination blade on the table saw (the inside corners are connected after glueup with a small chisel). The width of the groove is crucial and must be determined by trial and error with a piece of screen and dowel. The dowel must fit snugly and tightly enough to hold well, but not so tightly that the dowel can't be pressed below the surface of the wood. When the dowel fits correctly, the door is ready for glueup.

Gluing up—Before gluing up a door, I assemble it dry to make sure everything fits. I use two-part epoxy for my screen doors. Though it's expensive and has a relatively short pot life, it's waterproof, has excellent gap-filling ability and is transparent when it cures. Unlike most other glues, it actually bonds better on sawn surfaces than on planed wood. Also, epoxy can be mixed to be a bit

From *Fine Homebuilding* magazine (December 1989) 57:72-75

Wooden screen doors can be unobtrusive or ornamental, depending on the desired effect. The screen doors in the photos below are simple, allowing the entry doors to show through. The screen door pictured above, built of solid oak, conceals the main door and serves as an important visual element in the entryway.

The door below, built of Honduras mahogany, is simple and sturdy. The lock rail adds strength and divides the upper and lower half of the door behind it. The door pictured above is a stone's throw away from the ocean and features bronze screening, which resists corrosion. The storm doors over it protect it from nasty weather.

flexible so that it will move with the wood. Though epoxy usually requires a temperature of 65° F or above to cure, special formulas are available for use at colder temperatures. Epoxies have also been developed that cure quickly or adhere well to specific types of wood. The epoxy I use is made by Allied Resin Corporation (Weymouth Industrial Park, East Weymouth, Mass. 02189), but there are plenty of good brands available.

I use bar clamps and leave them on overnight. A long clamp tightened diagonally straightens the door if it's out of square.

If the door is to be finished, I finish it right after glueup, but before installing the screen. My doors are usually finished with an oil-base primer and paint, or with spar varnish with a UV filter.

Installing the screen—With the frame completed, the final step is installing the screen. There are several different types of screen on the market, though they aren't all easy to find. Most hardware stores carry aluminum and fiberglass screening, which are relatively inexpensive. Bright aluminum screening lets the most light through, but it dents, tears and corrodes easily. Charcoal-colored electro-alodized aluminum is tougher and resists corrosion. Fiberglass, the cheapest screen on the market, is easy to work with and won't dent. But it does stretch, and bluejays, grasshoppers and other critters like to chew on it.

I often use bronze screening. It's expensive and stiffer to work with than aluminum and

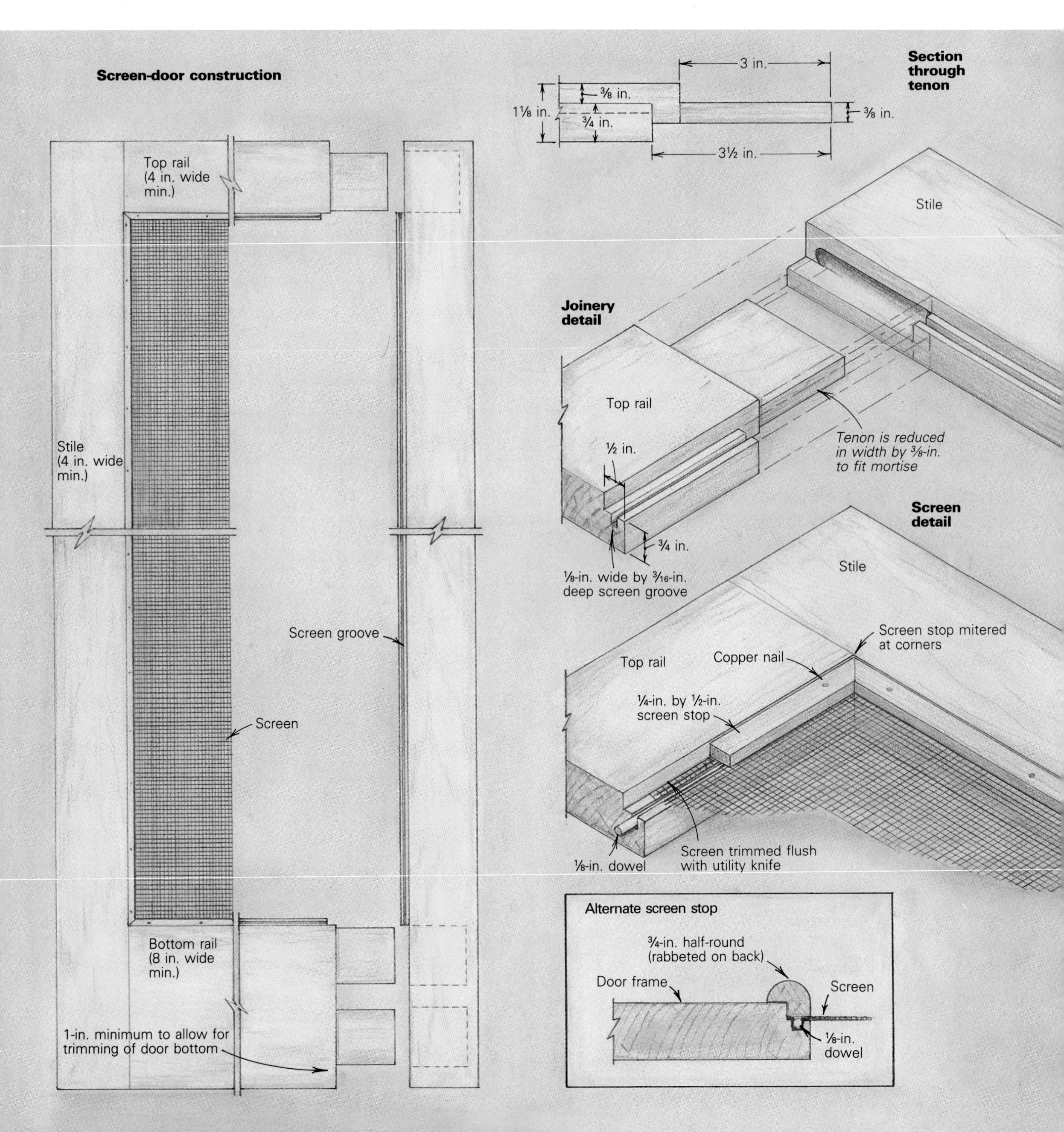

fiberglass, but it's strong and resistant to corrosion, an important consideration when installing screens in houses by the ocean (where I do most of my work). It also tarnishes to a greenish color, a look that some of my clients prefer. I buy mine at a local hardware store.

Screening is also made of galvanized steel (which turns chalky and disintegrates) and stainless steel (by far the most expensive and durable screen on the market). Some manufacturers even make a coated fiberglass screen called "solar shade" that blocks 70% of the sun's heat while providing ventilation. It's supposed to reduce air-conditioning bills and carpet fade. Hanover Wire Cloth, Inc. (E. Middle St., Hanover, Pa. 17331) calls their product Solar Guard or Solar View, depending on the mesh. The New York Wire Co. (152 N. Main St., Mt. Wolf, Pa. 17347) calls theirs Goldstrand Solar Screen.

Whatever the screening material, I use a splining tool to push it and the ⅛-in. dowel into the groove (photo below). A splining tool consists of a handle about 8-in. long with a narrow 1½-in. dia. metal wheel on either end. One wheel has a convex edge and is used to coax the screen into the groove. The other wheel has a concave edge and is used to press the dowel into the groove.

Because the tool is designed for installing screen in metal frames (where rubber gaskets are used instead of dowels to secure the screen), it doesn't fit the dowels quite right. But with practice and patience, it works fine. Some hardware stores sell splining tools, or they can be purchased from Elgar Products, Inc. (P. O. Box 22348, Cleveland, Ohio 44122).

I cut the screen 4 in. bigger than the opening, which leaves enough extra screen to grab onto and pull tight during installation. Starting at a stile, I press the screen into the groove with the convex wheel, making sure the weave of the screen is straight in relation to the frame. Then I turn the splining tool over and shove the dowel into the groove. This locks the screen into the groove.

Next, starting at the middle of the opposite side, I push the screen into the groove with the splining tool while pulling the screen tight with my other hand. The screen doesn't have to be perfectly tight because when I press the dowel into the groove, the screen is tightened further. After I install the second dowel, I repeat the process with the remaining two sides. When installed, the screen should be stretched nicely with no major depressions. If any objectionable dips remain, I pry out a dowel, tighten the screen and insert a new dowel.

The excess screen is bent into the corner of the rabbet with the splining tool and trimmed off at the corner with a razor knife. I cover the edge of the screen with a simple ¼-in. by ½-in. stop, mitered at the corners and fastened with copper nails (drawing facing page, lower right). For a more decorative effect, I sometimes use ¾-in. half-round instead, rabbeted on the back side so that it holds the screen and covers the edge of the door (drawing facing page, below right).

The screen is installed with the use of a splining tool. The tool consists of a handle with a metal wheel on either end, one with a convex edge and the other with a concave edge. To install the screen, the author presses the screen into the screen groove with the convex wheel. That done, he flips the tool over and presses a ⅛-in. dowel into the groove with the concave wheel (photo above), locking the screen into the groove. Excess screen is trimmed off at the corner of the rabbet with a utility knife.

Hardware—I use solid-brass hardware for its durability. Stanley Hardware (a division of The Stanley Works, New Britain, Conn. 06050) makes a 3-in. by 3-in. stamped solid-brass hinge with a ball tip that is well-made, reasonably priced and looks good.

For the door latch, I use a surface-mounted lockset made by Merit Metal Products Corp. (242 Valley Rd., Warrington, Pa. 18976). The lockset has a knob on the exterior side, a latch on the interior side and is lockable. As a bonus, it's easy to install. I'm extra careful, though, to install the lockset where it won't bump into the lockset on the primary door (I learned that lesson the hard way).

Storm doors—There isn't much difference between a screen door and a storm door, except that storm doors are designed to inhibit, rather than encourage, air infiltration. That means substituting ¼-in. thick laminated safety glass or tempered glass for the screen. Some screen-door manufacturers build doors with interchangeable panels to fill both functions. But the doors require additional hardware to secure the panels. Also, it's difficult to stretch screen tightly over a narrow, removable frame without causing the frame to bow. Plus, large panels of glass are difficult to handle in removable frames. I prefer to make two separate doors that can be interchanged by removing the hinge pins.

The only difference between my screen doors and storm doors is that I adjust the size of the rabbets to accommodate the glazing, and of course, I eliminate the screen grooves. The glass is contained in the rabbet with either glazing points and glazing compound or with wood stops. When using glazing compound, I cut the rabbets ¼ in. wide by ⅝ in. deep, which allows the proper slope for the glazing compound. The same size rabbet also works well in conjunction with ¼-in. quarter-round molding. Otherwise I cut the rabbets ¼ in. wide by ⅜ in. deep and use ¾-in. dia. half-round molding. The molding is rabbeted on the backside so that it laps over the edge of the door frame. □

Stephen Sewall of Sewall Associates, Inc., is a custom homebuilder in Portland, Maine. Photos by the author.

Ordering and Installing Prehung Doors

Precise results require careful planning

by Steve Kearns

Our crews consider installing anything but a prehung door on our jobs to be a throwback to the Dark Ages. With a few tricks and some basic carpentry skills, you can install a prehung door in half an hour or less. Sure, you pay more to have your doors prehung—our door shop reports the average price is $5 to $7 per door—but we find it unimaginable that anyone could save money or have better quality control by doing the hanging on site.

The walk-through—My first piece of advice is this: Don't order your doors off the plans. Details like door size, handing (hinges on the right or the left side) and the direction of swing can change as the walls go up. So we wait until all the door openings are roughed in before ordering the doors. There is an exception to this rule: We have our shop order the blanks for any custom doors that have a long lead time as soon as possible. The handing and the jamb widths may change, but the doors probably won't.

For interior doors, we frame our openings 2 in. wider and 2 in. taller than the door size. For example, a 2 ft. 6 in. by 6 ft. 8 in. door requires a 2 ft. 8 in. by 6 ft. 10 in. rough opening. The gaps allow for ¾-in. thick jambs and ¼ in. of shim space on each side. An exterior door comes with a threshold (photo above) and typically needs an additional 3 in. of height to allow for its thickness. Once the house is framed and the change orders accommodated, you know for certain the size of every door and the thickness of the walls.

Standing by. **Whether an interior or an exterior door, the pertinent information is written on the outside of the jamb. The notes for this exterior door include the contractor's name, the width of the jamb, the handing (L.H.), a code for the hinge type and the width and species of the threshold material. Three lines of Geocel caulk provide the weatherproof seal under the threshold.**

Now is the time to take a walk, plans in hand, with your door-shop representative to lay out and specify every door in the house. Our local shop, Pozzi of Idaho, has an order form for this purpose. The door specifications include its size, style, handing, jamb species and width, hinge size and finish, threshold (if needed), sweep, lockset and dead-bolt bores, type of hardware and any special details, such as the need for paint-grade or stain-grade jamb stock. If the form is filled out completely and accurately, you should be able to eliminate one bottle of aspirin from your list of supplies for the job. Making a mistake on the order form usually means you'll have to reorder or remodel something, then explain why to a long-faced client who showed up 10 minutes after the doors arrived.

Back to our walk-through. Our door man carries his order form, a pencil with an eraser, a carpenter's keel and a tape measure. I carry the blueprints. A good set of prints will have the doors numbered on the floor-plan pages and listed in a door "schedule." A comprehensive door schedule will include most of the information listed in the order form, but if you're trusting enough to order straight from the door schedule, keep the aspirin.

We start at the opening for door #1, measuring to check the called-out door against the framed opening. Assuming there's no difference, we mark the door's number and indicate its handing on the trimmer by drawing a rectangle with an *X* through it in the position of the top hinge. At the same time, our door man notes the rest of the pertinent specs for this door.

We move throughout the house in this manner until we have laid out every door. There should be agreement between the order form, the number of doors shown on the plan and the number listed in the schedule. If these numbers disagree, we reconcile the discrepancies.

Taking delivery—Now the day comes when the doors arrive on site. The scheduling is perfect: The drywall is hung, taped and textured; the walls are painted (we put masking tape over all the keel notes on the rough openings before the painter sprays the walls; otherwise, they will disappear); the floor is swept; and carpenters are waiting to take the doors off the truck. The shop has written each door's number on the outside of its jamb, and we take each door to its designated opening. We usually have someone reading the plans to direct traffic as the doors are unloaded. If the scheduling is off, we store the doors out of the way and under tarps until all the prep work is complete.

Installation—Once the doors are distributed, a carpenter begins installing each one by checking the basics: The floor should be level, the header

From *Fine Homebuilding* magazine (April 1992) 74:62-65

1. Flushing up. Carpenter Alex Harakay nails the strike-side jamb while checking that the exterior siding and the edge of the jamb remain flush.

2. Missing the shims. Pairs of nails are driven below the shims rather than through them, so the jamb can still be adjusted.

3. Fine-tuning the reveal. With the door closed, Harakay adjusts the shims to achieve an even space between the door and the jamb.

4. Bullnose corners. For a rounded corner detail, metal corner bead is tucked into factory-cut kerfs in the edges of the jambs.

square with the trimmers and the trimmers plumb in both directions.

If the floor is not level, you'll need to know which jamb leg needs to be adjusted to correct for it. Interior doors typically come with a 1¼-in. gap between the bottom of the door and the bottom of the jambs. That gap allows for a carpet pad and a carpet. If the flooring isn't as thick as a carpet, cut the bottoms of the jambs so that the door will clear the finished floor by ¼ in. to ⅜ in.

The hinge jamb is the first to be affixed to the framing. Ideally, the trimmer on the hinge side of the door should be straight and plumb, allowing the hinge jamb to be nailed to it without shims. Each jamb is affixed to its trimmer with three pairs of 16d finish nails. If the trimmers are slightly out of plumb in the plane of the wall, shimming the door jambs will be required.

If the opening is just a little out of plumb in the plane perpendicular to the wall, our rule is "go with the wall." That way the door casings will fit flush with both the wall and the jamb. If the trimmers are out of plumb in different directions perpendicular to the wall, we fix the wall before installing the door.

When a prehung door leaves our supplier, it has a couple of little shipping blocks under its jambs to protect the corners, as well as thin spacers between the jambs and the edges of the door, and several 6d duplex nails driven through the jamb into the door edges to keep everything aligned. The nails and the blocks have to be removed before the door is slid into its opening. If the hinge screws poke through the jamb, we cut them flush with a reciprocating saw so that the jamb can be tight to the trimmer.

Let's say our rough opening isn't perfect. Because we checked the hinge-side trimmer before we inserted the door, we know which way it's out of plumb and whether we should start nailing at the top or the bottom. In this example, the floor is level and we will need a shim near the bottom to plumb the jamb, so we start by nailing at the top of the jamb, just below the hinge. The door has to be open while the nails are driven home. We support the door in the open position with a stack of shims to keep it from pulling the jamb out of alignment.

A jamb for a typical 2x4 wall is 4⁹⁄₁₆ in. wide. That dimension equals the 3½ in. of the stud and two layers of ½-in. drywall plus ¹⁄₁₆ in. The extra width makes it easier to get the casings to lie flush on the jamb edges with no gaps. The jambs should be checked periodically for alignment with the walls as the nails are driven through the jambs into the trimmers.

Next, from each side we insert a pair of shims just below the bottom hinge and plumb the jamb with a 6-ft. 6-in. level. We "over-plumb" it slightly because nailing moves the trimmer away from the door opening a little. The nails should pass just below the shims, allowing the shims free movement for adjusting the jamb in and out. With the top and bottom of the jamb secured, we put our level on the jamb and shim the middle until it's flush with the level's edge. After setting the nails, we give the door another check for plumb. If necessary, a little tapping on the shims should get it perfect.

To secure the strike side, insert pairs of shims snugly so that they'll stay put at the top and bottom (top photos, p. 119). Secure the shims with nails driven just below them. Now close the door and adjust the shims in and out to get the correct reveal between the jamb and the door edge (bottom left photo, p. 119). We match the thickness of the spacer blocks at the top of the door.

The door is now plumb and level, and its jambs are flush with the drywall on each side. A key step (and one that is often left out) is to replace at least two of the top hinge screws and one from both the middle and bottom pairs with screws that are long enough to bite at least 1 in. into the trimmer. The trimmer, not the jamb, should hold up the door.

Fitting the stops—All that remains is setting the door stops, which come from the shop loosely attached to the jambs. For this task we'll need an example of the strike plate and the lockset that will be installed in the finished doors. We screw the strike plate into the precut mortise in the jamb and insert the latch bolt into its mortise in the edge of the door. The rest of the lockset is not necessary for this fitting. Now we close the door tightly to the front of the strike plate while nailing the stops in place with a finish nailer. If the doors and the jambs are to be painted rather than stained and lacquered, we insert a matchbook cover as a temporary spacer between the door and the stops to allow for the paint buildup.

This last step of temporarily installing the hardware may seem like overkill, but we have found it the only way to ensure that the finished doors will close firmly against the strike plate and the stops. The door is now ready for the casings to be installed by the trim carpenters.

The exception to the sequence I've just explained is the bullnose casing detail. A lot of the houses we've built lately have radiused corner beads, which require the jambs to be installed before the drywall. The corner bead fits into kerfed jambs (bottom right photo, p. 119). The door shop takes care of the kerfs, and we have them take off ¼ in. per side from the width of the jamb to allow the metal to wrap smoothly from the drywall to the jamb. For example, the jamb for the 2x6 wall shown here is 6¹⁄₁₆ in. wide.

Finishing the doors—Our painters usually take the doors off to finish them. And we have them number each door—inside the lockset bore is a good spot. They remove the hinges and the screws from both the door and the jamb and put the hardware in bags for safekeeping.

It takes longer to describe this process than to do it. When I do my estimate, I allocate an hour for installing each prehung door, including unloading and stocking them. My carpenters almost always beat the estimate.

Finally, store your painted doors until the carpet is laid. Now you're ready to go around in your socks and put each door back in place and install the locksets and strikes—all of which will fit perfectly because you ordered it so. □

Steve Kearns is a contractor based in Ketchum, Ida. Photos by Charles Miller.

Plumb-Bob Door Hanging

When the level won't work

By Scott Wynn

Few topics elicit more heated discussions among carpenters than the "right" way to hang doors. Now I'm going to throw another technique into the fray: hanging doors with a plumb bob.

On one of my early jobs as a carpenter, I had to hang about 25 doors in a remodeled Victorian. The new framing was poorly done. Many of the openings varied in width, and some tapered so much that there was zero clearance for shimming. The trimmers weren't even close to being plumb (in either direction). To make matters worse, as I nailed the jambs of the prehung doors to their lousy trimmers, the jambs began to twist, and they took on high spots between the shims that threw off my level readings. I'm sure there are ways to compensate for this, but I was having a hard time inventing them.

Then my boss came around. Perturbed at the slow progress, he asked, "Why don't you use a plumb bob?" That was the way he'd learned how to hang a door, and he was a little surprised that I wasn't using one already. Here, with some things I've learned since, is what he showed me.

Begin the job by removing the door from its hinges. Set the jambs in the opening and leave the stretcher on if you can. Check the floor for level. If it's out, temporarily shim the low side of the jamb until the head jamb is level. Transfer the distance you raised the jamb to the bottom of the other jamb and mark it. The amount you cut off the bottom of the jambs will depend on the finish flooring. I figure 1 in. for carpet and ¾ in. for hardwood.

Pull the frame out of the opening and trim the bottoms of the jambs as necessary. Remove the stretcher and the door stops and put the frame back in the opening. Now take a pair of shims (always try to use them in opposing pairs) and place them above the top hinge. In the portion of the jamb that will be covered by the stop, drive a 16d casing nail just below the shim—not through it—into the trimmer (drawings right).

Before you go any further, take a look at how the opposite jamb sits in relation to the corresponding wall. If the jamb is out of alignment, the cripple is probably twisted, and you will need to put in an odd shim to throw it back (drawing below right). A twisted trimmer usually requires an extra shim at each shim point.

Now tie the plumb-bob string to the nail so that the bob clears the jamb and falls just shy of the floor. I use a small plumb bob with about 8½ ft. of string tied to a short length of ½-in. dowel. I wind the unused line on the dowel and hang it out of the way over the ends of the top shims.

Place another pair of shims below the bottom hinges and secure them with a 16d nail driven below the shims and toward the hinge side of the string (to avoid hitting the plumb line). Shim and nail the

center hinge as well. You may have to add other pairs of shims, depending on the weight of the door, the number of hinges and the straightness of the jamb.

Now that the jamb is secured in several spots, note the distance from the string to the jamb where it is tied off (drawing below). It should be the same the entire length of the jamb, so drive the shims in or out to correct the distance to the string.

Once you've fine-tuned the shims to get the jamb perfectly plumb, drive in the 16d nails a bit more to make everything tight. Check everything again. When you are satisfied that the jamb is plumb, secure the shims with 8d (or larger) finishing nails. Now remove the string and drive in the 16d nails.

Rehang the door and shim the strike-side jamb and the head to an equal space around the door—usually 3/32 in. I like to use the same technique of placing a single 16d nail below the shims and then setting them with a couple of finishing nails. Put the top set of shims on the strike-side jamb near the head (drawing left). This placement will help counteract the weight of the door pulling the head jamb away from the hinge side of the door. Likewise, the jamb sometimes warps below the shims under the bottom hinge. Add shims near the floor to straighten it out.

Remind yourself not to set any shims or nails (including casing nails) around the strike-plate area. Trying to work around a protruding nail when mortising for a strike plate will slow you down considerably.

After installing the lockset and strike plate, reinstall the stops. Start with the strike-side stop. Set it tight to the door at the head and bottom but flex it away from the door a hair shy of 1/16 in. in the middle. This will allow the door to shut and latch crisply without it rebounding off the stop.

The hinge-side stop should be set about 1/32 in. away from the door so that the door edge won't bind on the stop as the door is closed. Allow an additional 1/32-in. minimum clearance for paint or varnish.

Now that I'm acquainted with its many uses, I always carry a plumb bob in my apron. It's easier to carry than a big level and a lot harder to knock over.

— Scott Wynn is an architect/contractor who also designs and builds furniture in San Francisco, Calif. Drawings by the author.

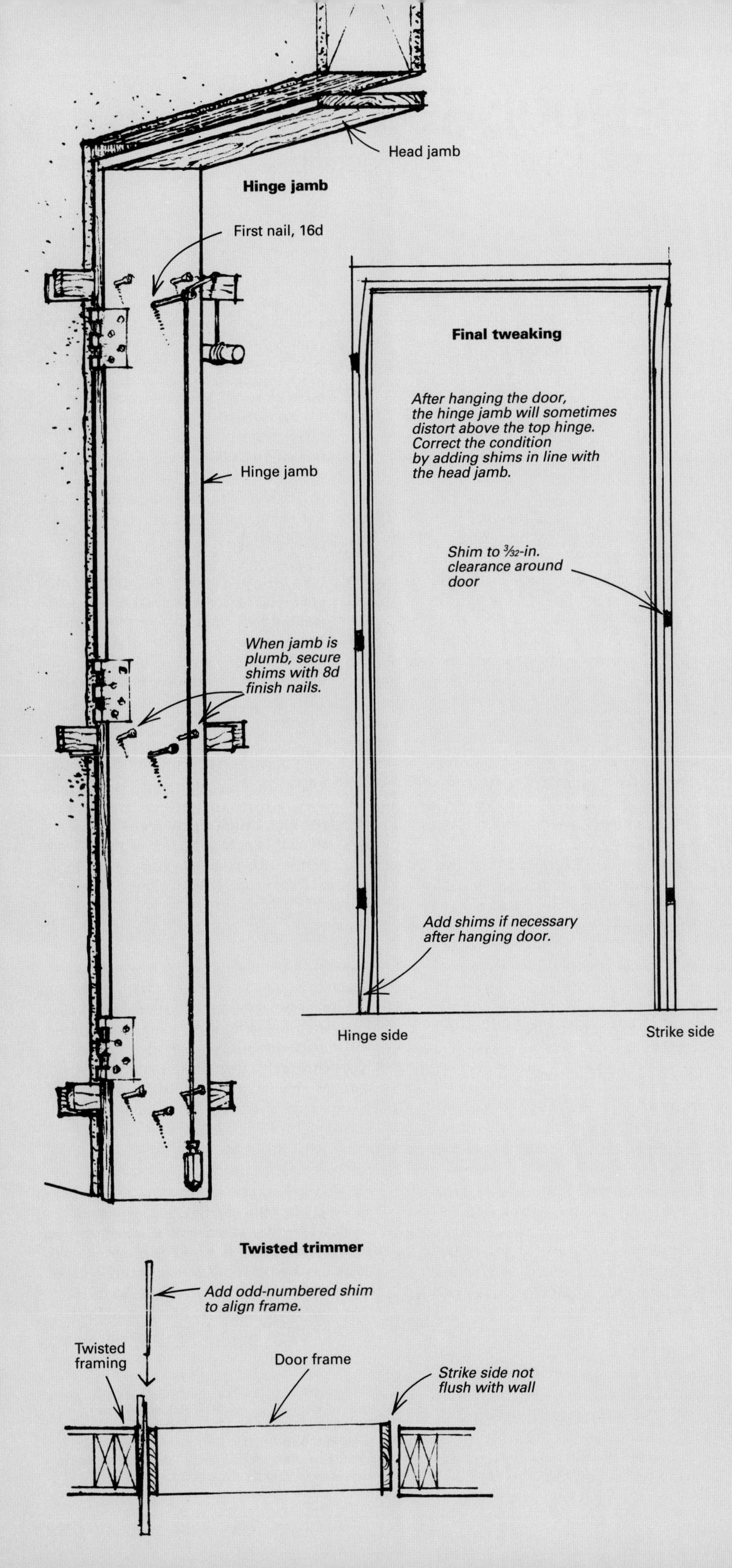

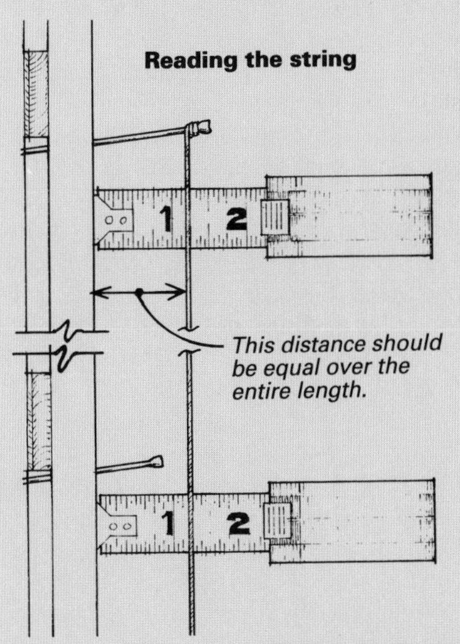

French-Door Retrofit

Opening up a wall to bring the outdoors in

by David Strawderman

What we know today as French doors were originally considered tall casement windows that simply reached to the floor. They appeared in the late 17th century at Versailles—Louis XIV's grandiose headquarters southwest of Paris. There they overlooked immense gardens that required the rerouting of a river for adequate irrigation.

French doors are still a gracious architectural element that can enhance the appreciation of a garden—no matter what its size. A pair of them make the connection more immediate than even a large window. My clients for the job illustrated in this article, Luis and Carol Fondavila, wanted a closer link between their breakfast/dining area and their beautifully landscaped backyard. Here's how I removed the existing double-hung window, created a 6-ft. wide opening and filled it with a pair of new French doors.

Headers and rough openings—Creating a new opening in a bearing wall requires that you shift the load from the existing studs to a post-and-beam carrier. The beam, or header, is typically a 4x timber or a couple of 2xs with a piece of ½-in. plywood sandwiched between. Then the entire assembly is spiked together with 16d nails. The depth of the header depends upon the span and load, and the common rule of thumb for single-story dwellings is that 1 in. of header depth will span 1 ft. For example, an 8-in. deep header will span 8 ft. A house with a typical 8-ft. ceiling and standard 6-ft. 8-in. doors will usually have enough space between the top plates and the door framing for a 12-in. deep header.

But what about two-story houses, like this one? I use common sense, look at what's in the wall and act accordingly. In this case, the existing 4-ft. wide window had a single 2x8 header. I decided that a double 2x8 would be adequate to span 6 ft. If you have any doubts about this kind of a calculation, however, have an engineer or architect size the header.

Another structural consideration besides the header depth is the foundation bearing capac-

French doors. **Traditional French doors consist of a pair of multi-light doors, both of which are operable. The primary door (in this pair, the one on the left) carries the lockset. The secondary door holds the strike plate, and is held fast by deadbolts.**

From *Fine Homebuilding* magazine (June 1991) 68:42-46

ity. The loads from the new header are passed by way of the trimmer studs to the foundation. This creates "point loads" on the foundation, and if they're considerable they can crack an otherwise adequate footing or a foundation atop weak soils. The Fondavilas' house has a massive foundation with 10-in. stemwalls in perfect condition. I was satisfied that the new loads from a 6-ft. long header weren't going to cause a problem. If the header span is longer than about 7 ft., however, I consult an engineer or an architect if I'm at all in doubt about the integrity of the structure. It's important that this be addressed *before* any holes are cut in the walls. Reinforcing the foundation at the post-bearing points can be messy, expensive and time-consuming, especially after the loads have already been altered.

Once I've considered the structural part of the equation, I turn my attention to utility obstructions. Telephone and television lines are minor items that are easily dealt with. Electrical wires, gas and water lines hidden in the wall can present more serious problems. First I check the exterior wall for utility entrances. If a gas line or an electric service is entering the house at precisely the spot where my client wants a new set of doors, I explain that the utility companies will have to relocate them, thus adding considerable expense. Perhaps the doors can be moved a little to accommodate the obstructions.

The interior wall offers clues to probable electrical wiring paths. Outlets typically have to be relocated, and light switches moved to the side of the new opening. Sometimes a look above the proposed door location, in the attic for example, will reveal evidence of wiring in the stud bays that will be affected. I also check in the basement or crawl space.

The most difficult utilities to relocate are gas, water and waste lines. Fortunately, most plumbing lines run through interior walls. If you are putting doors in a wall that already has a window, chances are you won't encounter plumbing lines in the portion of wall below the window. There are, of course, exceptions to this, such as long trap arms from a sink, or vent pipes that take a horizontal jog across a wall. Implanting a set of doors in a blank wall runs a greater risk of running into water and gas lines. To assess the likelihood of finding pipes, take a look in the crawl space or basement under the section of wall in question. Again, you may want to move a door a few feet this way or that to avoid relocating pipes. Upstairs bathrooms and vents in the roof can give you additional clues as you learn about where the pipes may be buried in the walls. Keep in mind that the more obstructions there are, the higher the cost will be. When I am satisfied that cutting and framing the rough opening is feasible, I address the design of the finished unit.

Style and costs—A pair of French doors engage each other in three basic ways: the hinged pair; a hinged door with a fixed sidelight panel; and a sliding pair. Deciding which kind to install is usually a matter of style.

For the classic look and a generous doorway that can be opened to the breezes, nothing compares with the traditional pair of hinged, true divided-light doors (photo facing page). Their delicate muntin grids are compatible with the windows of most homes built before the '50s. On the downside, a pair of hinged doors are tough to install and they are often considered the Achilles' heel of weatherization. It can be hard to keep wind-driven rain from getting through the gap between the doors, though properly installed compression weatherstripping can do the job. Another way to protect them from the weather is to install an awning-type canopy over them, such as the one designed by Bill Mastin (*FHB* #63, p. 56).

French doors with fixed sidelights are easier to weatherize, and I charge considerably less to install them because they have only one operable door. They can be a good solution when space is tight.

Sliding French doors can be very wide because they ride on rollers. Sliders aren't susceptible to being blown shut by the wind and they don't get in the way when they're open. Sliders without muntins are compatible with modern architectural styles.

The main floor of the Fondavilas' house, built in the '20s, already had several pairs of 7-ft. tall French doors. Even though they were no longer available off-the-shelf, my clients wished to duplicate them, as well as the hardware.

When I don't have to match existing doors, I buy standard units from my local supplier. There I can select from stock doors that have from 10 to 15 lights per door, and are either 2 ft. 6 in. or 3 ft. wide by 6 ft. 8 in. tall. These doors cost between $140 and $200 each. They are made of solid vertical-grain Douglas fir, and their single-glazed, tempered lights are double-bedded, which means they are glazed on both sides to help keep out the rain. I can buy the same doors prehung with jambs and a threshold for around $800 from local door shops, or pay about double that for some nationally known brands. For about $120 worth of material I can make my own jamb set and prehang the doors in less than half a day.

Doors that have to be made from scratch cost a lot more. One bid for the Fondavila doors came in at $650 apiece. I settled on a pair that cost $375 per door. Prehung custom door units rise in cost accordingly. In any case, locksets are extra.

If you order a custom set of prehung doors, your supplier will need to know the details of the doors, the vertical dimension from the bottom of the threshold to the top of the jamb, the horizontal dimension from the outside edges of the jambs, and the depth of the jamb from the finished interior wall to the finished exterior. The rough opening should allow a ¼-in. gap at each side and at the top.

Hardware—A pair of French doors may open out or in, and the door you open first is called the primary door (or the active leaf). The primary door holds the entry hardware, while the secondary door is secured to the upper jamb and threshold with sliding bolts. These bolts can be surface-mounted on the interior door side, such as those in the Fondavilas' house, or mortised into the door edge. The entry and deadbolt sets are the same as those used for single doors, and their strike plates are secured to the secondary door. The secondary door typically has nonoperable knobs to match the active leaf.

The closed doors need a stop where they meet in the middle. There are two basic solutions and the T-astragal is by far the most typical (top drawing, p.125). This molding strip is secured to one of the door edges. It's best to orient the crossbar of the T to the exterior

Out with the old. **A reciprocating saw makes short work of old plaster, studs and the nails that hold them together.**

Using a Carborundum masonry blade ensures a straight, clean cut in a stucco wall.

The new header is inserted into its cavity in the old wall. Note how the cripple studs over the window have been cut in a horizontal line so that they can bear equally on the header.

New trimmer studs support the weight of the header and its load, and define the edge of the rough opening. Here the door jambs are lifted into position and aligned with the interior wall.

side so that it will conceal the latchbolt and protect against the weather. The astragal may be on the secondary or primary door, depending on whether it swings in or out.

The alternative is a set of doors with interlocking, rabbeted edges. Although this configuration is elegant, the entry hardware for it is limited. The only company I know of that makes locksets and strikeplates suitable for rabbeted doors is Baldwin Hardware Corp. (P. O. Box 15048, Reading, Pa. 19612; 215-777-7811).

Jamb assembly—The jambs I use are made out of fir, and they are typically 1⅛ in. thick with a ½-in. deep rabbet along one edge to create an integral door stop. The interior edge of most jambs is flush with the interior finished wall, and the exterior edge is flush with the exterior wall. The width normally falls between 4¼ in. and 5¾ in. With the Fondavila job I had a 9½-in. deep wall, so I used 5¾-in. wide jambs and made up the difference with trim and stucco mold (drawing, p. 126).

A flat, open area is useful for layout and building the jamb set. I begin with the head jamb by marking off the dimensions of the doors, the astragal and its space and the hinge spaces (top drawing, facing page) on a piece of jamb stock. If I'm installing a threshold, its length is equal to the outside dimension of the assembled jamb plus the length of the ears. The side jamb fits into a tapered rabbet at the ends of the threshold, next to the ears (bottom drawing, facing page).

The length of the side jamb equals the door height, plus the thickness of the top jamb, the depth of the tapered rabbet in the threshold and a ⅛-in. gap above and below the door for clearance. If I'm installing a sweep or a gasket at the bottom of the door, I'll adjust that gap accordingly.

To join the jambs at the top corners, I cut a rabbet in the end of the side jambs and screw them to each end of the head jamb with three 3-in. drywall screws (bottom drawing). Before assembly, I lay down a bead of Polyseamseal caulk in the joint. The caulk serves as both a glue and a waterproofing agent (Polyseamseal, Darworth Co., 50 Tower La., Avon, Conn. 06001; 800-624-7767).

Incidentally, before I commit myself to a certain door height, I make sure the doors are square. They aren't always, and finding out after cutting the jambs is no fun.

Once I've assembled the jamb frame, I square it up and reinforce it with some diagonal 1x2s screwed into the edges of the jambs opposite the hinge side. The jamb frame can now be moved around pretty easily, and I lift it up onto the bench for easier access.

I typically install compression weatherstripping along the inside edge of the doorstops. I need to do this before the hinge gains are routed to allow space for the thickness of the weatherstripping. To cut the grooves for the weatherstripping, I use a slick little router made expressly for the purpose by Weatherbead Insualtion Systems Inc. (5321 Derry Ave. F, Agoura Hills, Calif. 91301; 800-966-0159). For more on this tool, refer to my review in *FHB* #60, p. 92.

While the jamb is on the bench, I bring the doors alongside so I can mortise the hinges. I use a Bosch 83038 Router Template (Robert Bosch Power Tool Corp., One Hundred Bosch Blvd., New Bern, N. C. 28562-4097; 919-636-4200) for this operation. With it, I can rout the gains for both the jambs and the doors at the same time. I use three hinges per door.

Next I affix the T-astragal to the correct door, and put a 3° bevel on the edge of the door that meets it. Because small misalignments are difficult to correct after installation, I wait until the unit is installed to position the lockset and security bolts.

If the client has decided that the doors should be painted instead of varnished or stained, now is the time to prepare them for priming by removing all putty and caulk around the glass. After sanding the doors and jambs with 120-grit sandpaper, I apply two coats of Kilz Oil Base Primer (Masterchem Industries Inc., P. O. Box 368, Barnhart, Mo. 63012; 800-325-3552).

Typical doors in plan

Sheathing
Threshold
Exterior casing
Threshold ears
3°
Secondary door
Primary door
⅛-in. hinge space
⅝-in. jamb
¼-in. shim space
1½-in. trimmer stud
1½-in. king stud
⅜-in. T-astragal stop
⅛-in. space

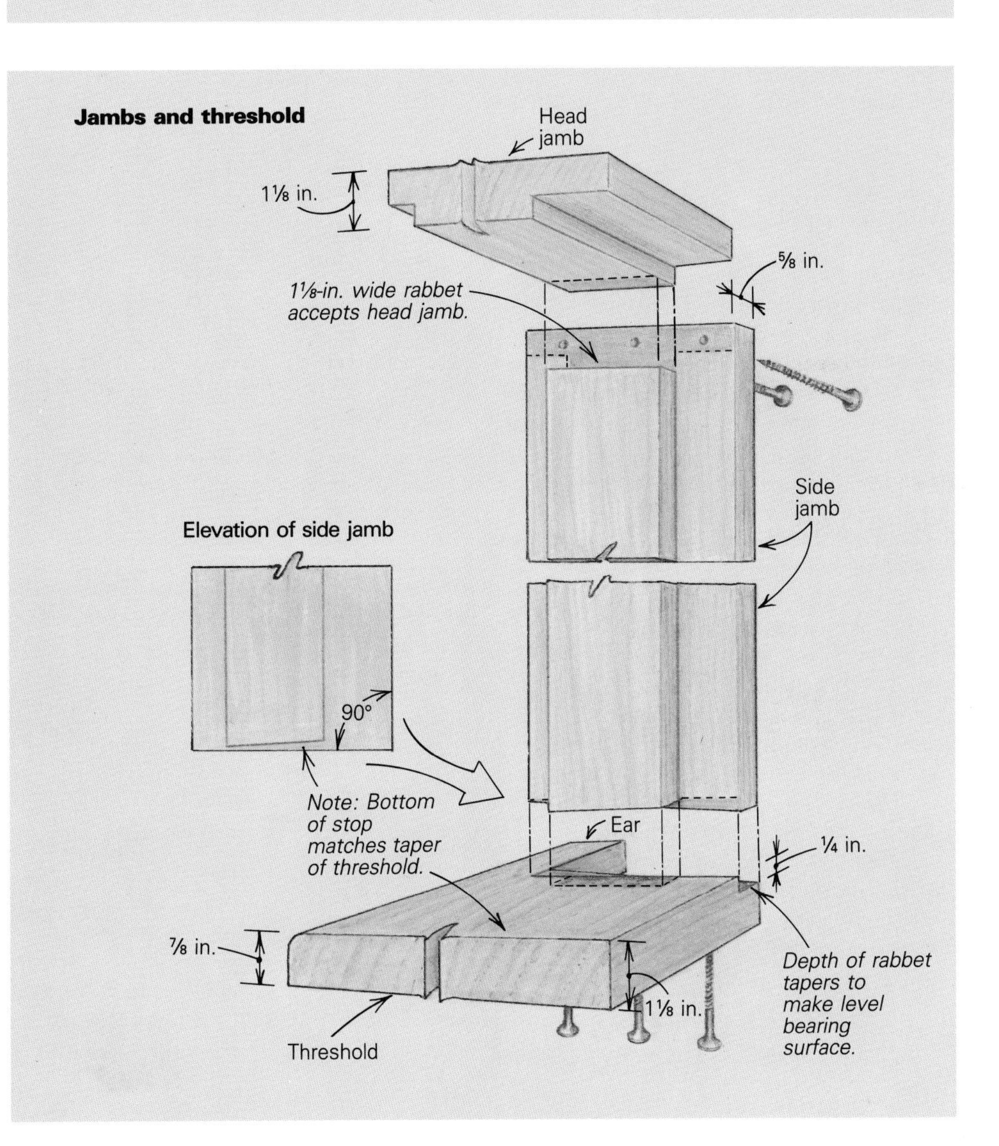

Opening up the wall—I begin work on the wall from the inside. This allows me to reposition any utilities, remove wallboard or plaster, and sometimes even assemble the new framing before breaching the exterior wall.

I locate the center of the new opening, and lay out the width of the door jambs, adding ¼ in. on each side. For an 8-ft. wall I draw vertical lines floor to ceiling on both sides. Then I use a reciprocating saw with a short plaster-cutting blade to cut the drywall or plaster. I hold the saw at a shallow angle to avoid hidden electrical wires (use a dust mask and goggles). If the vibration from the saw is cracking the plaster, I screw a 1x3 to the wall outside the cut line to hold the plaster together during the cut.

On an 8-ft. wall I remove the drywall or plaster all the way to the ceiling, and using a utility knife, score the inside corner where the wall meets the ceiling. At the Fondavilas' house, however, I stopped the cut at the point where the top of the new header would meet the old studs to avoid having to make a much larger patch in the 10-ft. wall (photo, p. 123). The wall's innards are now exposed, and if pipes or wires need to be moved, now is the time to do so.

Here's the typical sequence I follow when I'm making a rough opening 7 ft. wide or less. After removing the drywall or plaster, I use my reciprocating saw to cut in half the studs that need to be removed, and unless I find some evidence of concentrated loading from above (like a pinched sawblade as I crosscut a stud), I don't bother with shoring up the ceiling. A metal-cutting blade in the reciprocating saw makes it easy to cut the siding and sheathing

Drawings: Bob Goodfellow

nails away from the studs as I remove them.

Next I install the king studs on either side of the opening, slipping them into the wall cavity. I run a bead of Polyseamseal on the exterior edges of the new studs, and nail through the existing wall material into them after toenailing the studs to the top and bottom plates. The distance between the king studs should be 3½ in. more than the width of the finished unit. This distance equals the length of the new header.

In an upper corner, I make a notch in the drywall or plaster to allow easy insertion of the header. I tack the trimmer stud opposite the notch to its king stud, leaning its bottom out to allow ample clearance for the header. Then I hoist the header into place, put the trimmer under the other end of the header and nail it off. I tap the slanted trimmer stud into its final position and nail it securely to the king stud. If the new header doesn't reach all the way to the top plate, I put in blocks or cripple studs to restore the load path of the old studs. All the new and old framing members should be toenailed securely to one another.

If the opening is over 7 ft., I put up some temporary shoring a couple of feet in from the wall to help carry the weight of any ceiling joists that might be affected by the removal of the old studs. The shoring consists of a temporary 2x6 top and bottom plate the length of the new opening, along with some studs wedged in place and tacked to stay put while the new header is installed (for more on retrofitting headers, see *FHB #62*, pp. 85-87).

That's the theory, and often it follows that order. The Fondavila job had its exceptions. The height of the walls made it impossible to install new king studs without tearing out a lot more wall, so I made the header longer than it would normally be and turned the existing studs on either side of the opening into the king studs. I made a built-up header out of a pair 2x8s with 2x8 blocks between them to help fill the stud cavity and lifted the header into position (top right photo, p. 124). The new header is toenailed to the old studs, and it bears on the new trimmer studs that frame the rough opening (bottom photo, p.124).

Punching through—I begin the final process early in the day to ensure ample time to have a locked set of doors in place by evening. I begin by locating the corners of the opening on the outside of the house. A long ½-in. bit works nicely (I used a masonry bit for this job because of the stucco siding). Holes drilled, I move outside and snap chalklines for the cuts. They should be flush with the inside edge of the trimmer studs and the bottom edge of the header.

My helper, Larry Furniss, used a circular saw with a carborundum masonry blade to cut through the stucco finish (top left photo, p. 124). It's dirty work but cutting stucco this way ensures accuracy and leaves the remaining stucco undamaged. Then I used a reciprocating saw to sever the wood sheathing behind the stucco and any nails or studs that held this portion of the wall to the house.

Other exterior wall surfaces and molding details are cut differently. Cuts in horizontal siding, for example, sometimes need to be a few inches outside the rough opening to accommodate recessed molding. Bullnosed stucco-returns with no exterior wood molding need to be cut flush, then taken back several inches so new stucco wire can be tied in.

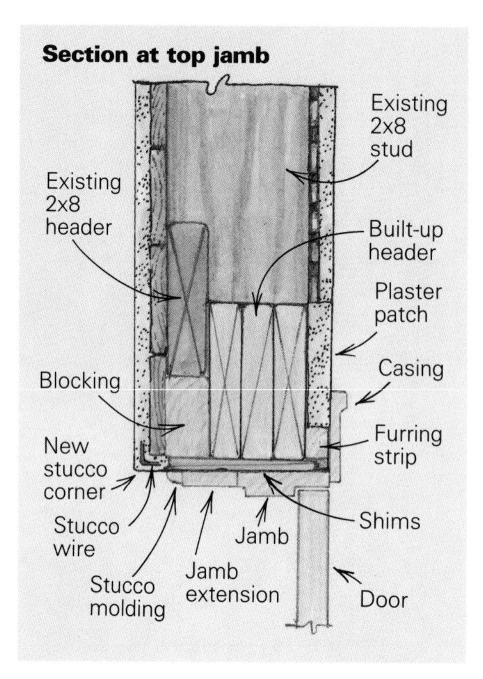

Installing the jambs—After removing the doors, Larry and I moved the jamb frame into place and lifted it into the opening from the outside until its interior edge was flush with the plane of the wall (bottom photo, p. 124). I didn't put a threshold on the Fondavilas' doors because they plan to have a mason create a cast-concrete sill and stoop to match the other doors. If the threshold is affixed to the jambs, however, its ears should be flush with the exterior finish. Check the jamb against the interior and exterior surfaces. Most walls have variations, so you'll have to average them out. When remodeling, I've found it better to conform to existing conditions than to follow my impulse and make everything plumb and level.

I wedge the jamb at the top to hold it in place without any nails while I hang the doors. Instead of leveling and squaring the frame, I use the doors as a guide to "squareness." If properly prehung, small in-and-out and up-and-down adjustments to the jambs will bring the doors into proper alignment (photo below). I shim the frame on each side 4 in. from the top and bottom, above and below the top and bottom hinges and near the center hinge. I shim the top jamb and threshold near each corner and at 2-ft. intervals. I used to secure the jambs with 16d coated finish nails, but have switched to countersunk 3-in. drywall screws. Once I've got the jambs anchored to the trimmers so that the doors operate properly, I install the lockset, deadbolt and surface bolts.

The jamb and doors are brought into square with one another by inserting shims and blocks between the trimmers, header and the jamb.

Touching up—I usually patch any holes that I've had to make in the interior walls with pieces of drywall. Because the Fondavilas have plaster-on-lath walls, however, I had my subcontractor use expanded metal lath affixed to the new header in order to anchor a plaster patch. Like many a plaster wall, these are kind of wavy, so I installed the new door casings and baseboards before the final coat of plaster. That allowed my plasterer to bring the finish coat right to the edge of the casings, filling up the gaps caused by the wavy walls.

On the outside, I sealed around the framing and jambs with foam insulation. Then I nailed on the exterior trim and stucco molding, and readied it for the stucco man with a couple of coats of primer.□

David Strawderman is a carpenter in Los Angeles, California. Photos by Charles Miller.

INDEX